INTRODUCTORY LAPLACE TRANSFORM WITH APPLICATIONS

ROHAN J. DALPATADU
G.S. SINGH
ASHOK K. SINGH

For Dr. G. S. Singh, in memoriam

Order this book online at www.trafford.com
or email orders@trafford.com

Most Trafford titles are also available at major online book retailers.

Print information available on the last page.

ISBN: 978-1-4907-6068-1 (sc)
ISBN: 978-1-4907-6069-8 (e)

Library of Congress Control Number: 2015910315

Trafford rev. 07/16/2015

Trafford PUBLISHING www.trafford.com
North America & international
toll-free: 1 888 232 4444 (USA & Canada)
fax: 812 355 4082

CONTENTS

PREFACE

Numerous books on theory and applications of the Laplace transform are available; most of these books are written primarily for mathematicians. By contrast, this book attempts to present the material in a simplified manner, and examples from various disciplines from existing literature are included to demonstrate the method of Laplace transform. A computer program in the R programming language for numerical inversion of the Laplace transform is included.

The target audience for this book is a researcher in one of the following disciplines: petroleum engineering, environmental engineering, contaminant hydrology, electrical engineering, finance and economics. This book can be used as supplemental material in undergraduate and graduate classes in engineering mathematics, and also as a handbook of Laplace transforms by practicing engineers and scientists.

1. INTRODUCTION

Integral transforms were introduced by Euler in 1763 as a tool for solving second-order linear differential equations, and Spitzer in 1878 gave the name of Laplace transform to the integral[1]

$$y = \int_a^b e^{sx}\phi(s)ds.$$

One of the first applications of the modern Laplace transform was by Bateman in 1910 who used it to transform Rutherford's equations in his work on radioactive decay[2]. The modeling of complex engineering and physical problems by linear differential equations has made the Laplace transform an indispensable mathematical tool for engineers and scientists. The method of Laplace transform for solving linear differential equations is very popular in the disciplines of electrical engineering, environmental engineering, hydrology, and petroleum engineering. The Laplace transform method is also used in statistics for deriving the distribution of sums of independent random variables. Applications of the Laplace transformcan be found in finance and economics. We briefly describe some applications of Laplace transforms in these disciplines.

Applications of the Laplace Transform in Electrical Engineering

The Laplace transform plays an important role in the discipline of control systems engineering. A control system is typically analyzed by computing the Laplace transforms of different functions of time, and then using inversion to find the solution to the problem. The method of Laplace transform is widely used in digital signal processing[3]as well.

1 Michael A. B. Deakin(1981). The development of the Laplace transform, 1737–1937. Archive for History of Exact Sciences, 30. XII. 1981, Vol. 25, pp 343-390.

2 Joel L. Schiff (1999). The Laplace Transform: Theory and Applications. Springer-Verlag, New York.

3 J M Blackledge (2006). Digital Signal Processing: Mathematical and Computational Methods, Software Development and Applications. Hardwood Publishing Limited, England.

Applications of the Laplace Transform in Environmental Engineering

Due to the complexity of the channels through which ground water flows, it is difficult to consider a model of the movement of ground water at a microscopic level. The French engineer Darcy, however, developed an equation that effectively averages the microscopic complexities, and provides a macroscopic model of ground water movement. Darcy's law is central to the derivation of equations used to model the flow of groundwater. The mathematical models, based on the physics of ground water flow and boundary conditions imposed by the ground water basin in question, usually take the form of a boundary value problem[4], which are solved either analytically or computationally[5]. Numerical techniques are the basis for digital computer simulation of transient ground water flow in aquifers. The two most widely used numerical methods for solving ground water equations are the finite difference and the finite element techniques.

Application of the Laplace Transform in Hydrology

The Laplace transform is commonly used in subsurface contaminant hydrology[6]. The problems of contaminant transport in fractured porous formation have attracted considerable attention especially when dealing with the disposal of radioactive waste in underground repositories[7]. Using Laplace transform techniques, numerousproblems arising in the study of migration of radionuclides in fractures as well as in the surrounding rock have been investigated. In the studies of radionuclear waste repositories, the problem of nuclide migration has been well investigated[8]. The Laplace transform has been used to solve the problem of migration of radionuclides in fissured rock under the influence of micropore

---

4 Mark M. Clark (2011). Transport Modeling for Environmental Engineers and Scientists. John Wiley & Sons.

5 Bear, J., 1972. Dynamics of Fluids in Porous Media, American Elsevier, New York, N.Y.

6 Rao S. Govindaraju (2007). Moment Analysis for Subsurface Hydrologic Applications. Springer.

7 Rao S. Govindaraju (2002). Stochastic Methods in Subsurface Contaminant Hydrology. SCE Publications.

8 Nuclear Fuel Safety Project (KBS), Handling of spent nuclear fuel and final storage of vitrified high-level reprocessing waste, vol. II, Geology, report, Stockholm, 1977

diffusion and longitudinal dispersion[9], the case of constant source strength[10], and for contaminant transport with radioactive decay in fractured porous rock under radial flow condition[11].

Analytical solutions for steady state tracer transport with precipitation-dissolution reactions in multiple-fracture systems have been obtained using the method of Laplace transform[12] .

Applications of the Laplace Transforms in Petroleum Engineering

The analysis of problems of steady state flow of fluid through sand formations dates back to 1897[13]. The problem of unsteady state radial flow of a compressible fluid in a porous media has been solved by transforming it into the heat equation.[14]The first application of the Laplace transform method in Petroleum Engineering was in solving unsteady state flow problems in oil reservoirs.[15]A majority of the problems treated in this article testify to the usefulness of the Laplace transform to solve difficult problems in reservoir engineering. It concerns itself primarily with the transient conditions prevailing in the oil reservoirs during the time they are produced. In investigations of effectiveness of hydraulic fracturing used to increase oil production, Laplace transform has been used to find solutions for finite conductivity models

9 Anders Rasmuson and IvarsNeretnieks (1981). Migration of Radionuclides in Fissured Rocks: The Influence of Micropore Diffusion and Longitudinal Diffusion. Jour. Geophys. Res., Vol. 86, pp. 3749-3758.

10 Anders Rasmuson (1984). Migration of Radionuclides in Fissured Rock: Analytical Solutions for the Case of Constant Source Strength. Water Resources Research, Vol. 20, pp. 1435-1442.

11 Chia-Shyun Chen (1986). Solutions for Radionuclide Transport from an Injection Well into a Single Fracture in a Porous Formation. Water Resources Research, Vol. 22, pp. 508-518.

12 Hui-Hai Liu, SumitMukhopadhyay, Nicolas Spycher, Burton M. Kennedy (2011). Analytic solutions of tracer transport in fractured rock associated with precipitation-dissolutionreactions. Hydrogeology Journal. Volume 19, pp 1151-1160.

13 Schlichter, U. S. G. S., Nineteenth Annual Report,. 1897-8.

14 William Hurst (1934). Unsteady Flow of Fluids in Oil Reservoirs. Journal of Applied Physics 5, 20-30.

15 Van Everdingen, A.F. and Hurst, W. (1949). The Application of the Laplace Transformation to Flow Problems in Reservoirs. Petroleum Transactions, AIME, Pp. 305-324.

assuming fractures to be rectangular[16] and elliptical cross sections[17]. Tracer tests are used to estimate heterogeneity and dispersivity, key parameters in the success or failure of enhances oil recovery or environmental remediation. The method of Laplace transform[18] and iterated Laplace transform[19] have been used to determine close form solutions to the tracer transport models in heterogeneous porous media.

In Chapter 2, we define the Laplace transform, present some of its properties, and then provide examples of deriving Laplace transforms and also computing the inverse Laplace transform. In Chapter 3, we present examples from engineering, statistics, and economics literature of how the method of Laplace transform is used to solve real problems. In Chapter 4, we present the method of numerical inversion of the Laplace transform, include a code in R[20] that numerically computes the inverse, and present several illustrative examples.

---

16 H. Cinco-Ley and F. Samaniego-V. "Transient Pressure Analysis for Fractured Wells," SOC. Pet. Eng. J. (Sept. 19Sl), 1749-1766.

17 Michael Francis Riley (1991). Finite Conductivity Fractures in Elliptical Coordinates. Ph.D. dissertation, Stanford University.

18 J.A. Barker (1982). Laplace transform solutions for solute transport in fissured aquifers, Adv. Water Resour. 5, 98–104.

19 Ibrahim Kocabas (2011). Application of iterated Laplace transformation to tracer transients in heterogeneous porous media. Journal of the Franklin Institute, Volume 348, pp. 1339–1362.

20 R Core Team (2014). R: A language and environment for statistical computing. R Foundation for Statistical Computing, Vienna, Austria. URL http://www.R-project.org/.

2. THE LAPLACE TRANSFORM AND ITS PROPERTIES

Definition 2.1: Let $f:[0,\infty)\to\mathbb{R}$. The Laplace transform $\mathcal{L}\{f\}$ of a function *f* is given by:

$$\mathcal{L}\{f\}=F(s)=\int_0^\infty e^{-st}f(t)dt. \tag{2.1}$$

Theorem 2.1: (Existence Theorem). Let $f:[0,\infty)\to\mathbb{R}$ be piecewise continuous on $[0,b)$ for each $b>0$. If there exist constants *M* and γ such that

$$|f(t)|\le Me^{\gamma t} \text{ for all } t\ge 0 \tag{2.2}$$

then the Laplace transform exists for all $s>\gamma$.

Proof: The function $e^{-st}f(t)$ is integrable over any interval of the form $[0,n]$ since f is piecewise continuous. Therefore, for any $b>0$ and $s\ne\gamma$,

$$\left|\int_0^b e^{-st}f(t)dt\right|\le\int_0^b e^{-st}|f(t)|dt\le\int_0^b e^{-st}Me^{\gamma t}dt$$
$$=\frac{M}{s-\gamma}[1-e^{-(s-\gamma)b}]\le\frac{M}{s-\gamma} \text{ if } s>\gamma.$$

Note: The existence condition (2.2) is sufficient for the Laplace transform to exist, but may not be necessary.

Theorem 2.2 (Linearity): If $\mathcal{L}\{f\}$ and $\mathcal{L}\{g\}$ are the Laplace transforms of f and g, respectively, then

$$\mathcal{L}\{af+bg\}=a\mathcal{L}\{f\}+b\mathcal{L}\{g\},$$

where *a* and *b* are constants.

Proof:

$$\mathcal{L}\{af+bg\}=\int_0^\infty [af(t)+bg(t)]e^{-st}dt$$
$$=a\int_0^\infty e^{-st}f(t)dt+b\int_0^\infty e^{-st}g(t)dt$$
$$=a\mathcal{L}\{f\}+b\mathcal{L}\{g\}.$$

Theorem 2.3 (The Laplace Transform of a Derivative): Let $f:[0,\infty)\to\mathbb{R}$ be continuous and satisfy (2.2) for some constants M and γ. If f is differentiable on $[0,\infty)$ and f' is piecewise continuous on $[0,b]$ for any $b>0$, then the Laplace transform of f' exists for all $s>\gamma$ and

$$\mathcal{L}\{f'\}=s\mathcal{L}\{f\}-f(0).$$

Proof: Suppose f' is continuous on $[0,\infty)$. Then for $s>\gamma$

$$\int_0^b e^{-st}f'(t)dt= e^{-st}f(t)\Big|_0^b+s\int_0^b e^{-st}f(t)dt$$

$$= e^{-sb}f(b)-f(0)+s\int_0^b e^{-st}f(t)dt$$

$$= s\int_0^b e^{-st}f(t)dt-f(0).$$

As $b\to\infty$,

$$e^{-sb}f(b)\to 0 \text{ and } \int_0^b e^{-st}f(t)dt \to \mathcal{L}\{f\}$$

since $|f(t)|\le Me^{\gamma t}$. Hence result for the case when f' is continuous.

Suppose f' is not continuous. Choose $b>0$ such that at least one of the discontinuities of f' is contained in the interval $[0,b]$. Let $t_1<t_2<\ldots<t_n$ be the discontinuities of f' in the interval $[0,b]$. Then for $s>\gamma$

$$\int_0^b e^{-st}f'(t)dt = \int_0^{t_1} e^{-st}f'(t)dt + \sum_{k=1}^{n-1}\int_{t_k}^{t_{k+1}} e^{-st}f'(t)dt + \int_{t_n}^{b} e^{-st}f'(t)dt$$
$$= e^{-st}f(t)\Big|_0^{t_1} + s\int_0^{t_1} e^{-st}f(t)dt$$
$$+\sum_{k=1}^{n-1}\left[e^{-st}f(t)\Big|_{t_k}^{t_{k+1}} + s\int_{t_k}^{t_{k+1}} e^{-st}f(t)dt\right]$$
$$+e^{-st}f(t)\Big|_{t_n}^{b} + s\int_{t_n}^{b} e^{-st}f(t)dt$$
$$= -f(0) + e^{-sb}f(b) + s\int_0^b e^{-st}f(t)dt$$
$$\to s\mathcal{L}\{f\} - f(0) \text{ as } b\to\infty.$$

This proves the result for the case when f' is not continuous.

Corollary 1: Let $f,\ f':[0,\infty)\to\mathbb{R}$ be continuous and satisfy (2.2) for some constants M and γ. If f' is differentiable on $[0,\infty)$ and f'' is piecewise continuous on $[0,b]$ for any $b>0$, then the Laplace transform of f'' exists for all $s>\gamma$ and

$$\mathcal{L}\{f''\} = s^2\mathcal{L}\{f\} - sf(0) - f'(0).$$

Proof: $\mathcal{L}\{f''\} = s\mathcal{L}\{f'\} - f'(0)$

$$= s\left[s\mathcal{L}\{f\} - f(0)\right] - f'(0)$$
$$= s^2\mathcal{L}\{f\} - sf(0) - f'(0).$$

Corollary 2: Let $f,\ f', f'',\dots, f^{(n-1)}:[0,\infty)\to\mathbb{R}$ be continuous and satisfy (2.2) for some constants M and γ. If $f^{(n-1)}$ is differentiable on $[0,\infty)$ and $f^{(n)}$ is piecewise continuous on $[0,b]$ for any $b>0$, then the Laplace transform of $f^{(n)}$ exists for all $s>\gamma$ and

$$\mathcal{L}\{f^{(n)}\} = s^n\mathcal{L}\{f\} - s^{n-1}f(0) - s^{n-2}f'(0) - \dots - f^{(n-1)}(0).$$

Proof: This is easily proven by mathematical induction.

Example 1: Let $f(t) = e^{at}$. Then

$$\mathcal{L}\{f\} = \int_0^\infty e^{-st} e^{at} dt$$
$$= \int_0^\infty e^{-(s-a)t} dt$$
$$= -\frac{e^{-(s-a)t}}{s-a}\Big|_0^\infty$$
$$= \frac{1}{s-a},\ \textit{for } s > a.$$

Example 2: Let $f(t) = 1$. By Example 1,

$$\mathcal{L}\{f\} = \mathcal{L}\{e^{0t}\}$$
$$= \frac{1}{s-0},\ \textit{for } s > 0.$$
$$= \frac{1}{s},\ \textit{for } s > 0.$$

Example 3: Let $f(t) = t^n$, $n \in \mathbb{N}$. Then

$$\mathcal{L}\{f\} = \int_0^\infty e^{-st} t^n dt$$
$$= -t^n \frac{e^{-st}}{s}\Big|_0^\infty - \int_0^\infty -\frac{e^{-st}}{s} n t^{n-1} dt$$
$$= \frac{n}{s}\mathcal{L}\{t^{n-1}\},\ \textit{for } s > 0.$$

$$\mathcal{L}\{t^n\} = \frac{n}{s}\mathcal{L}\{t^{n-1}\}$$
$$= \frac{n}{s}\frac{n-1}{s}\mathcal{L}\{t^{n-2}\}$$
$$.$$
$$.$$
$$= \frac{n}{s}\frac{n-1}{s}\cdots\frac{2}{s}\frac{1}{s}\mathcal{L}\{t^0\}$$
$$= \frac{n!}{s^n}\mathcal{L}\{1\}$$
$$= \frac{n!}{s^n}\frac{1}{s} = \frac{n!}{s^{n+1}},\ \text{for } s > 0.$$

Example 4: Let $f(t)=\cosh(at),\ a>0$. Then

$$f(t)=\frac{1}{2}e^{at}+\frac{1}{2}e^{-at}.$$

Therefore,

$$\begin{aligned}\mathcal{L}\{f\}&=\frac{1}{2}\frac{1}{s-a}+\frac{1}{2}\frac{1}{s+a},\quad \text{for } s>a.\\ &=\frac{s}{s^2+a^2},\quad \text{for } s>a.\end{aligned}$$

Similarly, $\mathcal{L}\{\sinh(at)\}=\dfrac{a}{s^2+a^2},\quad \text{for } s>a.$

Example 5: Let $f(t)=\cos(bt),\ b>0.$ Then $f'(t)=-b\sin(bt)$ and $f''(t)=-\cos(bt)$. The functions $f,\ f',$ and f'' are continuous on $[0,\infty)$. Furthermore,

$$\begin{aligned}&|f(t)|\le 1=1\cdot e^{0t},\ \text{for } t\ge 0,\\ &|f'(t)|\le b=b\cdot e^{0t},\ \text{for } t\ge 0.\end{aligned}$$

If $M=\max\{1,\ b\}$, then

$$|f(t)|\le Me^{0t} \text{ and } |f'(t)|\le Me^{0t},\ \text{for } t\ge 0.$$

Therefore, by Corollary 1 of Theorem 2.3,

$$\begin{aligned}&\mathcal{L}\{-b^2\cos bt\}=s^2\mathcal{L}\{\cos bt\}-s\cos(0)-\sin(0)\\ &-b^2\mathcal{L}\{\cos bt\}=s^2\mathcal{L}\{\cos bt\}-s.\end{aligned}$$

Hence, $\mathcal{L}\{\cos bt\}=\dfrac{s}{s^2+b^2},\ b>0.$

Similarly, $\mathcal{L}\{\sin bt\}=\dfrac{b}{s^2+b^2},\ b>0.$

Note: The Gamma function is defined as

$$\Gamma(x)=\int_0^\infty t^{x-1}e^{-t}dt,\quad x>0.$$

If x is a positive integer (say n), then

$$\Gamma(n)=\int_0^\infty t^{n-1}e^{-t}dt=(n-1)!.$$

This follows because

$$\Gamma(n)=(n-1)\Gamma(n-1)$$

if n is a positive integer greater than 1 and

$$\Gamma(1)=\int_0^\infty 1\cdot e^{-t}dt=1.$$

Example 6: Let $f(t)=t^\alpha,\ \alpha\in(0,\infty)$. Then

$$\begin{aligned}\mathcal{L}\{f\}&=\int_0^\infty e^{-st}t^\alpha dt\\&=\int_0^\infty e^{-\tau}\left(\frac{\tau}{s}\right)^\alpha\frac{d\tau}{s},\text{ where }\tau=st\text{ and }s>0\\&=\frac{1}{s^{\alpha+1}}\int_0^\infty \tau^{(\alpha+1)-1}e^{-\tau}d\tau\\&=\frac{\Gamma(\alpha+1)}{s^{\alpha+1}},\ s>0.\end{aligned}$$

Note: If α is a positive integer (say n), then $\mathcal{L}\{f\}=\dfrac{\Gamma(n+1)}{s^{n+1}}=\dfrac{n!}{s^{n+1}},\ s>0$ as expected.

Theorem 2.5(Shifting): If the Laplace transform of f is $F(s)$ for $s>\gamma$, then the Laplace transform of $e^{at}f(t)$ is $F(s-a)$ for $s>a+\gamma$.

Proof: By definition

$$\mathcal{L}\{e^{at}f(t)\}=\int_0^\infty e^{-st}e^{at}f(t)dt=\int_0^\infty e^{-(s-a)t}f(t)dt=F(s-a).$$

Example 7:

(a) $$\mathcal{L}\{t^n\}=\frac{2}{s^{n+1}}\text{ if }n\in\mathbb{N}.$$

Therefore,

$$\mathcal{L}\{t^n e^{at}\}=\frac{2}{(s-a)^{n+1}}.$$

(b) $$\mathcal{L}\{\sin bt\} = \frac{b}{s^2 + b^2}.$$

Therefore,

$$\mathcal{L}\{e^{at}\sin bt\} = \frac{b}{(s-a)^2 + b^2}.$$

Definition 2.2: The inverse Laplace transform of a function $F(s)$, $s > \gamma$ for some $\gamma \in \mathbb{R}$, denoted by $\mathcal{L}^{-1}\{F\}$, is the function *f* given by

$$f(t) = \frac{1}{2\pi i}\int_{c-i\infty}^{c+i\infty} e^{ts}F(s)ds,$$

where *c* is a real number.

Note: If $F = \mathcal{L}\{f\}$, then it can be shown that $\mathcal{L}^{-1}\{F\} = f$.

Theorem 2.6 (The Laplace Transform of an Integral): Let $f:[0,\infty) \to \mathbb{R}$ be piecewise continuous and satisfy (2.2) for some constants *M* and *γ*. Then the Laplace transform of the function defined by $\int_0^t f(\tau)d\tau$ exists for $s > \max\{0, \gamma\}$ and

$$\mathcal{L}\left\{\int_0^t f(\tau)d\tau\right\} = \frac{1}{s}\mathcal{L}\{f\}.$$

Proof: Suppose $\gamma \le 0$. Then

$$|f(t)| \le Me^{\gamma t} \le Me^{-\gamma t}.$$

Therefore, we may assume that $\gamma > 0$. The function defined by $\int_0^t f(\tau)d\tau$ is continuous on $[0,\infty)$ and

$$\left|\int_0^t f(\tau)d\tau\right| \le \int_0^t |f(\tau)|d\tau \le \int_0^t Me^{\gamma\tau}d\tau$$

$$= \frac{M}{\gamma}\left[e^{\gamma t} - 1\right] \le M'e^{\gamma t} \quad \text{for all } t \ge 0$$

where $M' = M/\gamma$.

Furthermore, $\frac{d}{dt}\left(\int_0^t f(\tau)d\tau\right)=f(t)$ except for discontinuities of f. Therefore, $\frac{d}{dt}\left(\int_0^t f(\tau)d\tau\right)$ is piecewise continuous on $[0,b]$ for any $b>0$, and by Theorem 2.3

$$\mathcal{L}\{f\}=s\mathcal{L}\left\{\int_0^t f(\tau)d\tau\right\}-\int_0^0 f(\tau)d\tau.$$

Hence result.

Corollary: Let $f:[0,\infty)\to\mathbb{R}$ be piecewise continuous and satisfy (2.2) for some constants M and γ. Then

$$\mathcal{L}^{-1}\left\{\frac{1}{s}F(s)\right\}=\int_0^t f(\tau)d\tau.$$

Proof: Follows immediately from Theorem 2.5 and Definition 2.2.

Example 8: Find the inverse Laplace transform of $F(s)=\frac{5}{(s-2)^4}+\frac{2s-3}{s^2-9}$.

$$F(s)=\frac{5}{(s-2)^4}+\frac{2s-5}{s^2-9}$$

$$=\frac{5}{(s-2)^4}+\frac{2s}{s^2-3^2}-\frac{5}{s^2-3^2}$$

$$=5e^{2t}\frac{t^3}{3!}+2\cosh 3t-\frac{5}{3}\sinh 3t$$

$$=\frac{5}{6}t^3e^{2t}+2\cosh 3t-\frac{5}{3}\sinh 3t.$$

Example 9: Find the inverse Laplace transform of $F(s)=\frac{2s-3}{s^2+4s+13}$.

$$F(s)=\frac{2s-3}{s^2+4s+13}.$$

Therefore,

$$\mathcal{L}^{-1}\{F\}=f(t)=2e^{-2t}\cos 3t-\frac{7}{2}e^{-2t}\sin 3t.$$

$$=\frac{2s-3}{(s+2)^2+3^2}$$

$$= \frac{2(s+2)-7}{(s+2)^2+3^2}$$

$$= \frac{2(s+2)}{(s+2)^2+3^2} + \frac{-7}{(s+2)^2+3^2}.$$

Example 10: Find the inverse Laplace transform of $F(s) = \frac{s^2+5}{s^3-3s-2}$.

We use the method of partial fractions:

$$F(s) = \frac{s^2+5}{s^3-3s-2}$$

$$= \frac{s^2+5}{(s+1)^2(s-2)} \equiv \frac{A}{(s+1)} + \frac{B}{(s+1)^2} + \frac{C}{(s-2)}$$

$$\Rightarrow s^2+5 \equiv A(s+1)(s-2) + B(s-2) + C(s+1)^2$$

$$\Rightarrow A = 0,\ B = -2,\ \text{and } C = 1.$$

$$F(s) = -\frac{2}{(s+1)^2} + \frac{1}{(s-2)}.$$

Therefore,

$$\mathcal{L}^{-1}\{F\} = f(t) = -2te^{-t} + e^{2t}.$$

Example 11: Find the inverse Laplace transform of $F(s) = \frac{2}{s^2(s^2+1)}$.

Instead of using partial fractions, we can use the corollary to Theorem 2.6.

First we will find $\mathcal{L}^{-1}\left\{\frac{2}{s(s^2+1)}\right\}$.

$$\mathcal{L}^{-1}\left\{\frac{2}{s(s^2+1)}\right\} = \int_0^t 2\sin\tau d\tau$$

$$= 2 - 2\cos t.$$

Therefore,

$$\mathcal{L}^{-1}\left\{\frac{2}{s^2(s^2+1)}\right\} = \mathcal{L}^{-1}\left\{\frac{1}{s}\cdot\frac{2}{s(s^2+1)}\right\}$$

$$= \int_0^t (2-2\cos\tau)d\tau$$

$$= 2t - 2\sin t.$$

Example 12: Find the solution of the initial value problem:

$$y'' - y' - 2y = 6\cos 2t - 2\sin 2t + 2;\ y(0) = 1,\ y'(0) = 0.$$

We take Laplace transforms to obtain:

$$\left(s^2Y - s\cdot 1 - 0\right) - \left(sY - 1\right) - 2Y = \frac{6s}{s^2+4} - \frac{4}{s^2+4} + \frac{2}{s} = \frac{8s^2 - 4s + 8}{s(s^2+4)},$$

where $Y = \mathcal{L}\{y\}$.

This reduces to:

$$Y = \frac{s-1}{s^2 - s - 2} + \frac{8s^2 - 4s + 8}{s(s^2+4)(s-2)(s+1)}$$
$$= \frac{s^4 - s^3 + 12s^2 - 8s + 8}{s(s^2+4)(s-2)(s+1)}.$$

Now using partial fractions:

$$Y = \frac{s^4 - s^3 + 12s^2 - 8s + 8}{s(s^2+4)(s-2)(s+1)} \equiv \frac{A}{s} + \frac{Bs + C}{s^2+4} + \frac{D}{s-2} + \frac{E}{s+1}$$
$$\Rightarrow A = -1,\ B = -1,\ C = 0,\ D = 1, \text{ and } E = 2.$$

Therefore,

$$Y = \frac{-1}{s} + \frac{s}{s^2+4} + \frac{1}{s-2} + \frac{2}{s+1}$$

and the solution is:

$$y = -1 + \cos 2t + e^{2t} + 2e^{-t}.$$

Theorem 2.7 (The Laplace Transform of a Convolution): Suppose $f, g : [0,\infty) \to \mathbb{R}$ are piecewise continuous and satisfy (2.2) for some constants M and γ. The convolution of f and g, denoted by $f * g$ is: $f * g = g * f = \int_0^t f(\tau)g(t-\tau)d\tau$. Then $\mathcal{L}\{f * g\} = F(s)G(s),\ s > \gamma$.

Proof: We change the order of integration in the double integral to obtain

$$\begin{aligned}
\mathcal{L}\{f*g\} &= \int_0^\infty e^{-st}\left(\int_0^t f(\tau)g(t-\tau)d\tau\right)dt \\
&= \int_0^\infty f(\tau)\left(\int_\tau^\infty e^{-st}g(t-\tau)dt\right)d\tau \\
&= \int_0^\infty f(\tau)\left(\int_\sigma^\infty e^{-s(\sigma+\tau)}g(\sigma)d\sigma\right)d\tau \\
&= \int_0^\infty f(\tau)\left(e^{-s\tau}G(s)\right)d\tau \\
&= F(s)G(s).
\end{aligned}$$

Example 13: Find the inverse Laplace transform of $H(s)=\dfrac{2}{s^2(s^2+1)}$.

We use the above theorem to find the inverse transform. Let

$$F(s)=\frac{1}{s^2} \text{ and } G(s)=\frac{2}{s^2+1}.$$

Then

$$\begin{aligned}
&f(t)=t \text{ and } g(t)=2\sin t \\
&\mathcal{L}^{-1}\{H\}=\mathcal{L}^{-1}\{F(s)G(s)\} \\
&\quad =f*g \\
&\quad =\int_0^t \tau 2\sin(t-\tau)d\tau \\
&\quad =-2\int_0^t \tau\sin(\tau-t)d\tau \\
&\quad =2\tau\cos(\tau-t)\Big|_0^t-2\int_0^t\cos(\tau-t)d\tau \\
&\quad =2t-0-2\sin(\tau-t)\Big|_0^t \\
&\quad =2t-2\sin t.
\end{aligned}$$

Example 14: Consider the second order linear differential equation with constant coefficients:

$$ay''+by'+cy=f,$$

where a, b, and c are constants and f is a function that satisfies (2.2) for some constants M and γ. Let the initial conditions be: $y(0)=y_0$ and $y'(0)=y_1$. Take Laplace transforms to obtain:

$$\begin{aligned}
&a(s^2Y-sy_0-y_1)+b(sY-y_0)+cY=F(s) \\
&Y(s)=\frac{ay_0s-ay_1-by_0}{as^2+bs+c}+\frac{1}{as^2+bs+c}F(s).
\end{aligned}$$

Let $G(s)=\dfrac{ay_0s-ay_1-by_0}{as^2+bs+c}$ and $H(s)=\dfrac{1}{as^2+bs+c}$. Use partial fractions to find the functions g and h. Then

$$Y(s)=G(s)+H(s)F(s)$$
$$y=g+h*f.$$

Example 15: Find a general solution of the second order equation: $y''-3y'+2y=2t-1$.

We will first assume that the initial conditions are the same as that of the previous example. Then

$$G(s)=\frac{y_0s-y_1+3y_0}{s^2-3s+2}$$
$$=\frac{A}{s-1}+\frac{B}{s-2}$$
$$\Rightarrow A=y_1-4y_0 \text{ and } B=5y_0-y_1.$$

$$H(s)=\frac{1}{s^2-3s+2}$$
$$=\frac{1}{s-2}-\frac{1}{s-1}.$$

Therefore,

$$g(t)=Ae^t+Be^{2t} \text{ and } h(t)=e^t-e^{2t}.$$

$$(h*f)(t)=\int_0^t(2\tau-1)(e^{2(t-\tau)}-e^{t-\tau})d\tau$$

$$=(2\tau-1)\left(-\frac{e^{2(t-\tau)}}{2}+e^{t-\tau}\right)\Bigg|_0^t-\int_0^t\left(-\frac{e^{2(t-\tau)}}{2}+e^{t-\tau}\right)\cdot 2d\tau$$

$$=(2t-1)\left(-\frac{e^0}{2}+e^0\right)+1\left(-\frac{e^{2t}}{2}+e^t\right)-\left(\frac{e^{2(t-\tau)}}{2}-2e^{t-\tau}\right)\Bigg|_0^t$$

$$=t-\frac{1}{2}-\frac{e^{2t}}{2}+e^t-\left(\frac{e^0}{2}-2e^0\right)+\left(\frac{e^{2t}}{2}-2e^t\right)$$

$$=t+1-e^t.$$

Therefore, a general solution is given by:

$$y(t) = Ae^{t} + Be^{2t} + t + 1 - e^{t} = A_1 e^{t} + Be^{2t} + t + 1,$$

where A_1 and B are arbitrary.

Theorem 2.8: Let $f:[0,\infty) \to \mathbb{R}$ be piecewise continuous and satisfy (2.2) for some constants M and γ, and $n \in \mathbb{N}$. Then $\mathcal{L}\{t^n f(t)\} = (-1)^n F^{(n)}(s),\ s > \max\{0,\gamma\}$.

Proof:

$$\begin{aligned}\frac{d}{ds}F(s) &= \frac{d}{ds}\int_0^\infty e^{-st} f(t)dt\\ &= \int_0^\infty -te^{-st} f(t)dt\\ &= (-1)\mathcal{L}\{tf(t)\}.\end{aligned}$$

Therefore,

$$\frac{d^2}{ds^2}F(s) = \frac{d}{ds}\left(\frac{d}{ds}F(s)\right)$$

$$\begin{aligned}&= \frac{d}{ds}\left((-1)\mathcal{L}\{tf(t)\}\right) = (-1)(-1)\mathcal{L}\{t(tf(t))\}\\ &= (-1)^2\mathcal{L}\{t^2 f(t)\}\end{aligned}$$

.

.

$$\frac{d^n}{ds^n}F(s) = (-1)^n\mathcal{L}\{t^n f(t)\}.$$

Hence result.

Example 15: Find the Laplace transform of $g(t) = te^{-2t}\cos 3t$.

Let

$$f(t) = e^{-2t}\cos 3t.$$

Then

$$F(s) = \frac{s+2}{(s+2)^2 + 3^2}$$

and

$$G(s)=(-1)^1\frac{d}{ds}\left(\frac{s+2}{s^2+4s+13}\right)$$

$$=-\frac{1.(s^2+4s+13)-(s+2)(2s+4)}{(s^2+4s+13)^2}=\frac{s^2-5}{(s^2+4s+13)^2}.$$

Definition 2.3: The unit step function is defined as:

$$U(t-t_0)=\begin{cases}0 & \text{if } t<t_0\\ 1 & \text{if } t\geq t_0\end{cases},\ t_0\geq 0.$$

Example 16: Let $f(t)=U(t-t_0)$. Then

$$\mathcal{L}\{f\}=\int_0^\infty U(t-t_0)dt$$
$$=\int_0^{t_0}e^{-st}\cdot 0dt+\int_{t_0}^\infty e^{-st}\cdot 1dt$$
$$=\frac{e^{-t_0 s}}{s},\ s>0.$$

Note: If $t_0=0$, then the unit step function is simply the function given by $f(t)=1$, whose Laplace transform is:

$$\mathcal{L}\{f\}=\frac{e^{-s\cdot 0}}{s}=\frac{1}{s},\ s>0.$$

Theorem 2.8: Suppose $\mathcal{L}\{f\}=F(s),\ s>0$. Then $\mathcal{L}\{f(t-h)U(t-h)\}=e^{-hs}F(s),\ s>0$, where $h\geq 0$.

Proof:

$$\mathcal{L}\{f(t-h)U(t-h)\}=\int_0^\infty e^{-st}f(t-h)U(t-h)dt$$
$$=\int_h^\infty e^{-st}f(t-h)\cdot 1dt$$
$$=\int_0^\infty e^{-s(\tau+h)}f(\tau)d\tau=e^{-hs}F(s).$$

Example 17: Find the Laplace transform of $f(t)=\begin{cases}\sin t & \text{if } 0\leq t<\pi\\ 1+\cos t & \text{if } t\geq\pi.\end{cases}$

We can write the function *f* as:

$$f(t) = \left(1 - U(t-\pi)\right)\sin t + U(t-\pi)(1-\cos t)$$
$$= \sin t + \sin(t-\pi)U(t-\pi) + U(t-\pi) + \cos(t-\pi)U(t-\pi).$$

Therefore,

$$F(s) = \frac{1}{s^2+1} + \frac{e^{-\pi s}}{s^2+1} + \frac{e^{-\pi s}}{s} + \frac{se^{-\pi s}}{s^2+1}.$$

Note: If $f(t)$ is defined for $0 \le a \le t < b$, then $f(t) = \left(U(t-a) - U(t-b)\right)f(t)$. The function f can be written as $f\left((t-h)+h\right)$ and expanded with an argument of $(t-h)$, where $h = a$ or b. Furthermore, $U(t-0) = 1$ and $U(t-\infty) = 0$ for all $t \ge 0$.

Theorem 2.9: Suppose $f:[0,\infty) \to \mathbb{R}$ is piecewise continuous and satisfies (2.2) for some constants M and γ and f is periodic with period T. Then

$$\mathcal{L}\{f\} = \frac{1}{1-e^{Ts}} \int_0^T e^{-st} f(t)dt, \; s > 0.$$

Proof:

$$F(s) = \int_0^T e^{-st} f(t)dt + \int_T^\infty e^{-st} f(t)dt = \int_0^T e^{-st} f(t)dt + \int_0^\infty e^{-s(\tau+T)} f(\tau+T)d\tau$$

$$= \int_0^T e^{-st} f(t)dt + e^{-Ts}\int_0^\infty e^{-s\tau} f(\tau)d\tau = \int_0^T e^{-st} f(t)dt + e^{-Ts}F(s)$$

$$\Rightarrow (1-e^{-Ts})F(s) = \int_0^T e^{-st} f(t)dt.$$

The result follows.

Note: If we define the function f_T as:

$$f_T(t) = \begin{cases} f(t) & \text{if } 0 \le t < T \\ 0 & \text{if } t \ge T, \end{cases}$$

then

$$\mathcal{L}\{f\} = \frac{1}{1-e^{-Ts}} F_T(s), \; s > 0.$$

Example 18: Find the Laplace transform of the square wave function given by

$$f(t) = \begin{cases} h, & 0 < t < c \\ -h, & c < t < 2c \end{cases}$$

and

$$f(t+2c)=f(t) \text{ for all } t\in\mathbb{R}.$$

Define the function f_{2c} as:

$$f_{2c}(t)=\begin{cases} h, & 0<t<c \\ -h, & c<t<2c \\ 0, & \text{otherwise} \end{cases}$$

Then

$$f_{2c}(t)=h(1-U(t-c))-h(U(t-c)-U(t-2c))=h-2hU(t-c)+h(U(t-2c)$$

and

$$F_{2c}(s)=\frac{1}{s}-2\frac{e^{-cs}}{s}+\frac{e^{-2cs}}{s}=\frac{h}{s}(1-e^{-cs})^2.$$

Thus the Laplace transform of f is given by

$$\mathcal{L}\{f\}=\frac{1}{1-e^{-Ts}}F_T(s),\ T=2c$$
$$=\frac{h(1-e^{-cs})^2}{s(1-e^{-2cs})}=\frac{h(1-e^{-cs})}{s(1+e^{-cs})}=\frac{h}{s}\tanh\frac{cs}{2}.$$

Definition 2.3: The Dirac delta function is defined as:

$$\delta(t-t_0)=\begin{cases} 0 & \text{if } t\neq t_0 \\ \text{undefined} & \text{if } t=t_0 \end{cases}$$

along with the property

$$\int_{-\infty}^{\infty}\delta(t-t_0)dt=1.$$

This is a generalized function and the integral is a Lebesgue integral. It also has the property

$$\int_{-\infty}^{\infty}f(t)\delta(t-t_0)dt=f(t_0),$$

where f is any integrable function. Furthermore,

$$\int_{-\infty}^{t}\delta(t-t_0)dt=U(t-t_0),\ t_0\geq 0.$$

Note: The Dirac delta function can be thought of as an impulse (function). It is quite useful in engineering applications.

Example 19: The Laplace transform of the Dirac delta function is:

$$\mathcal{L}\{\delta(t-t_0)\} = \int_0^\infty e^{-st}\delta(t-t_0)dt$$
$$= e^{-t_0 s}.$$

Example 20: A particle initially at rest is subject to an impulse of magnitude 2 at time $t=3$. The motion of the particle is governed by the differential equation:

$$y'' + 4y = 2\delta(t-\pi).$$

Since the particle was at rest initially, $y(0) = y'(0) = 0$. Taking Laplace transforms of both sides of the equation, we obtain

$$s^2Y + 4Y = e^{-\pi s}.$$

Therefore,

$$Y(s) = \frac{2e^{-\pi s}}{s^2+4}$$

$$y(t) = \sin 2(t-\pi)U(t-\pi) = \begin{cases} 0 & \text{if } 0 \le s < \pi \\ \sin 2t & \text{if } s \ge \pi \end{cases}$$

Example 21: The total distance travelled by a particle at time t is given by: $y(t) = 2 - e^{-2t}$. The particle is then subject to an impulse of magnitude 2 at time $t = 3$. Find the total distance travelled by the particle as a function of t.

The differential equation for the motion of the particle is:

$$y'' + 2y' = 2\delta(t-3)$$

with initial conditions: $y(0) = 1$, $y'(0) = 2$, because $y' = 2e^{-2t}$ and $y'' = -4e^{-2t}$. Taking Laplace transforms, we obtain

$$(s^2Y - s\cdot 1 - 2) + 2(sY - 1) = 2e^{-3s}$$

$$Y(s) = \frac{s+4}{s^2+2s} + \frac{2e^{-3s}}{s^2+2s} = \frac{2}{s} - \frac{1}{s+2} + \frac{e^{-3s}}{s} - \frac{e^{-3s}}{s+2}$$

$$y(t)=2-e^{-2t}+\left(1-e^{-2(t-3)}\right)U(t-3)=\begin{cases}2-e^{-2t} & \text{if } 0\le t<3\\ 3-(1+e^{6})e^{-2t} & \text{if } t\ge 3\end{cases}$$

Definition 2.4: Let $f=f(x,t)$ from $\mathbb{R}\times[0,\infty)\to\mathbb{R}$ be continuous and satisfy (2.2) in the second variable t, for fixed x and some constants M and γ. The (partial) Laplace transform of f with respect to the second variable t is defined by:

$$\mathcal{L}_t\{f(x,t)\}=F(x,s)=\int_0^{\infty} e^{-st}f(x,t)dt.$$

Theorem 2.10: Let $f=f(x,t)$ from $\mathbb{R}\times[0,\infty)\to\mathbb{R}$ be continuous and satisfy (2.2) in the second variable t, for fixed x and some constants M and γ. Suppose the partial derivatives $\frac{\partial f}{\partial t}$ and $\frac{\partial^2 f}{\partial t^2}$ are continuous and satisfy (2.2) in the second variable t, for fixed x and the same constants M and γ. Then

$$\mathcal{L}_t\left\{\frac{\partial f}{\partial t}(x,t)\right\}=sF(x,s)-f(x,0)$$

and

$$\mathcal{L}_t\left\{\frac{\partial^2 f}{\partial t^2}(x,t)\right\}=s^2F(x,s)-sf(x,0)-\frac{\partial f}{\partial t}(x,0).$$

Proof: Follows immediately from Theorem 2.3 and its corollaries.

Theorem 2.11: Let $f=f(x,t)$ from $\mathbb{R}\times[0,\infty)\to\mathbb{R}$ be continuous and satisfy (2.2) in the second variable t, for fixed x and some constants M and γ. Suppose the partial derivatives $\frac{\partial f}{\partial x}$ and $\frac{\partial^2 f}{\partial x^2}$ are continuous and satisfy (2.2) in the second variable t, for fixed x and the same constants M and γ. Then

$$\mathcal{L}_t\left\{\frac{\partial f}{\partial x}(x,t)\right\}=\frac{\partial F}{\partial x}(x,s)$$

and

$$\mathcal{L}_t\left\{\frac{\partial^2 f}{\partial x^2}(x,t)\right\}=\frac{\partial^2 F}{\partial x^2}(x,s).$$

Proof: $$\mathcal{L}_t\left\{\frac{\partial f}{\partial x}(x,t)\right\}=\int_0^\infty e^{-st}\frac{\partial f}{\partial x}(x,t)dt=\frac{\partial}{\partial x}\int_0^\infty e^{-st}f(x,t)dt=\frac{\partial F}{\partial x}(x,s).$$

$$\mathcal{L}_t\left\{\frac{\partial^2 f}{\partial x^2}(x,t)\right\}=\int_0^\infty e^{-st}\frac{\partial^2 f}{\partial x^2}(x,t)dt=\frac{\partial^2}{\partial x^2}\int_0^\infty e^{-st}f(x,t)dt=\frac{\partial^2 F}{\partial x^2}(x,s).$$

Example 22 (The Vibrating String): Consider a stretched string of finite length L held fixed at both ends and given an initial displacement of $\lambda\sin\frac{\pi x}{L}$ and zero initial velocity. The associated differential equation for the subsequent motion is:

$$\frac{\partial^2 u}{\partial t^2}=c^2\frac{\partial^2 u}{\partial x^2},\quad 0\le x\le L \text{ and } t\ge 0.$$

The initial conditions are:

$$u(x,0)=\lambda\sin\frac{\pi x}{L} \text{ and } \frac{\partial u}{\partial t}(x,0)=0.$$

The boundary conditions are:

$$u(0,t)=u(L,t)=0.$$

Taking Laplace transforms of the differential equation and the boundary conditions, we obtain:

$$c^2\frac{d^2U}{dx^2}=s^2U-s\lambda\sin\frac{\pi x}{L}$$

and

$$U(0,s)=U(L,s)=0.$$

The general solution of this equation is:

$$U(x,s)=C_1(s)e^{sx/c}+C_2(s)e^{-sx/c}+\frac{s\lambda\sin(\pi x/L)}{s^2+c^2\pi^2/L^2}.$$

The transformed boundary conditions give us:

$$C_1(s)+C_2(s)=0 \text{ and } C_1(s)e^{sL/c}+C_2(s)e^{-sL/c}=0.$$

Therefore,

$$C_1(s) = C_2(s) = 0$$

and

$$U(x,s) = \frac{s\lambda}{s^2 + c^2\pi^2 / L^2} \sin\frac{\pi x}{L}.$$

The inverse transform of the above is the solution to our problem:

$$u(x,t) = \lambda \cos\frac{c\pi t}{L} \sin\frac{\pi x}{L}.$$

3. APPLICATIONS OF THE LAPLACE TRANSFORM

In this chapter, the method of Laplace transform for solving differential equations is illustrated by several examples from several disciplines. In Examples 3.1 to3.4, the flow of contaminant of tracer through porous media is considered. Examples 3.5 to3.9 are concerned with modeling of radionuclide transport through porous media. In each of these examples, analytical solutions are obtained for different boundary conditions, and under simplifying conditions. Environmental and petroleum engineers and hydrologist routinely solve flow problems to predict the movement of reservoir fluids to screen effective injection schemes, or to design efficient monitoring networks. Analytical solutions such as the ones presented in this chapter provide means to assess the accuracy of a numerical scheme for solving more complicated contaminant transport problems.

Example 3.1

The first application of the Laplace transform in petroleum engineering[21] was in solving unsteady state flow problems in reservoirs. Assuming that the fluid flow is governed by Darcy's law, which relates the volume *v* of fluid with viscosity μ, flowing per unit of time through each unit of a porous medium such as sand with permeability K to the pressure gradient at the radial distance r:

$$v = \frac{K}{\mu}\frac{\partial P}{\partial r} \tag{3.1}$$

A material balance on a concentric element yields the equation of continuity for the radial system

$$\frac{K}{\mu}\frac{\partial\left(\rho r\frac{\partial P}{\partial r}\right)}{\partial r} = fr\frac{\partial P}{\partial T} \tag{3.2}$$

[21] Van Everdingen, A.F. and Hurst, W. (1949). The Application of the Laplace Transformation to Flow Problems in Reservoirs. Petroleum Transactions, AIME, Pp. 305-324.

where f is the porosity of the formation.

It is known that fluid density ρ decreases with pressure P; assuming that the rate of decrease is exponential, i.e., $\rho = \rho_0 e^{-c(P_0 - P)}$, differentiation gives

$$\frac{1}{\rho c}\frac{\partial \rho}{\partial r} = \frac{\partial P}{\partial r}$$

Substituting the above in (3.2) yields

$$\left(\frac{\partial^2 \rho}{\partial r^2} + \frac{1}{r}\frac{\partial \rho}{\partial r}\right)\frac{K}{f\mu c} = \frac{\partial \rho}{\partial T} \tag{3.3}$$

For liquids which are only slightly compressible, $\rho \approx \rho_0\left[1 - c(P_0 - P)\right]$, $P < P_0$.

The equations (3.2) and (3.3) lead to the following equation for the flow model:

$$\frac{\partial^2 P}{\partial r^2} + \frac{1}{r}\frac{\partial P}{\partial r} = \frac{\partial P}{\partial t} \tag{3.4}$$

where $t = \dfrac{KT}{f\mu c R_b^2}$

r = radial distance as a multiple of R_b, the well radius in centimeters

$K =$ permeability in darcys

$T =$ time in seconds

$f =$ prorsity, a fraction

$\mu =$ viscosity in centipoises

$c =$ compressibility

Applying Laplace transform to (3.1.4), we get

$$\frac{d^2 \bar{P}}{dr^2} + \frac{1}{r}\frac{d\bar{P}}{dr} = p\bar{P} \tag{3.5}$$

where $\bar{P} = \int_0^\infty e^{-pt} P(t)dt$ is the Laplace transformation of P.

The general solution for equation (3.1.5) is

$$\bar{P}(r,p) = AI_0\left(r\sqrt{p}\right) + BK_0\left(r\sqrt{p}\right) \tag{3.6}$$

where $I_0\left(r\sqrt{p}\right)$ and $K_0\left(r\sqrt{p}\right)$ are modified Bessel functions of order zero of the first and second kind respectively, and A and B are constants to be determines from the initial and boundary conditions.

The Constant Rate Case

The boundary conditions for this case for an infinite medium are:

(1) $P(r,t) \equiv 0$ at every point in the formation, and

(2) $\left(\dfrac{\partial P}{\partial r}\right)_{r=1} = -1$ at all times.

The modified Bessel functions are known to have the following behavior[22]:

For $x \ll n$,

$$I_n(x) \approx \frac{1}{n!}\left(\frac{x}{2}\right)^n, \; n \geq 0$$

$$K_0(x) \approx -\ln(x)$$

$$K_n(x) \approx \frac{(n-1)!}{2}\left(\frac{x}{2}\right)^{-n}, \; n > 0,$$

and for $x \gg n$,

$$I_n(x) \approx \frac{1}{\sqrt{2\pi x}}\exp(x)$$

$$K_n(x) \approx \frac{\pi}{\sqrt{2\pi x}}\exp(-x)$$

To obey the first initial condition, Ain (3.6) must be zero. Hence

$$\overline{P}(r,p) = BK_0\left(r\sqrt{p}\right) \tag{3.7}$$

The second boundary condition is

$$\left(\frac{\partial P}{\partial r}\right)_{r=1} = -1$$

Using the fact that the Laplace transform of 1 is $\dfrac{1}{p}$,

---

22 William H. Press, Saul A. Teukolsky, William T. Wetterling, and Brian P. Flanner (1992). Numerical Recipes in C++, The Art of Scientific Computing, Second Edition. Cambridge University Press., p. 236.

$$\left(\frac{\partial \bar{P}}{\partial r}\right)_{r=1} = \frac{-1}{p}.$$

Differentiating (3.1.6), the second boundary condition becomes

$$\left(\frac{\partial \bar{P}}{\partial r}\right)_{r=1} = B\sqrt{p}K'_0\left(\sqrt{p}\right) = -B\sqrt{p}K_1\left(\sqrt{p}\right).$$

Therefore

$$B = \frac{1}{p^{3/2}K_1\left(\sqrt{p}\right)}.$$

Substituting the above value of the constant *B* in (3.1.7), we get

$$\bar{P}(r,p) = \frac{K_0\left(r\sqrt{p}\right)}{p^{3/2}K_1\left(\sqrt{p}\right)} \tag{3.8}$$

To invert the Laplace transform (3.8) for large *p* or small *t*, we use the fact that

$$K_n(z) = \sqrt{\frac{\pi}{2z}}e^{-z} \tag{3.9}$$

and obtain

$$\bar{P}(1,p) = p^{-3/2} \tag{3.10}$$

The inverse of (3.10) is

$$P(t) = \frac{2}{\sqrt{\pi}}t^{1/2} \tag{3.11}$$

which implies that for small $t = KT/f\mu cR_b^2$, or large boundary radius R_b, the pressure drop for the unit rate of production approximates the linear flow condition.
When p is small or t is large, (3.8) is approximately

$$\bar{P}(1,p) = -\frac{\log p}{2p} + \frac{(\log 2 - \gamma)}{p}. \tag{3.12}$$

The inverse of the Laplace transform (3.12)is

$$\begin{aligned} P &= \frac{1}{2}(\log 4t - \gamma) \\ &= \frac{1}{2}[\log t + 0{\cdot}80907]. \end{aligned} \tag{3.13}$$

Equation (3.13) gives the solution for the continuous point source problem for large t. This is applicable for the study of interference between flowing wells[23],[24].

---

23 W. A. Bruce (1943). Pressure Prediction for Oil Reservoirs, Trans., AIME.

24 Lincoln F. Elkins (1946). Reservoir Performance and Well Spacing. Drilling and Production Practice, API-109.

Example 3.2

The case of a thin rigid fracture situated in a saturated porous rock is considered[25]. Following assumptions about the physical system are made-

(1) The groundwater velocity in the fracture is constant.
(2) A contaminant source of constant strength is present at the origin of the fracture.
(3) The width of the fracture is negligible in comparison to its length.
(4) There is complete mixing across the fracture width at all times, due to transverse diffusion and dispersion.
(5) Transport in the rock matrix is mainly due to molecular diffusion.
(6) Transport along the fracture is much faster than that within the host rock.
(7) Adsorption on the fracture surface is governed by a linear equilibrium isotherm $s=f(c)$.

Let

z = coordinate along the fracture axis,
t = time,
$c = c(z,t)$ = concentration of solute in solution,
s= mass of solute absorbed per unit length of fracture,
$2b$ = fracture width,
v = groundwater velocity,
λ = decay constant,
$t_{1/2}$ = half-life of the contaminant,
q = diffusive flux perpendicular to the fracture axis,
R = retardation coefficient = $1 + (K_f/b)$
K_f = distribution coefficient of the surface
D = hydrodynamic dispersion coefficient = $\alpha_L v+D^*$ with
α_L= dispersivity in the fracture axis direction, and
D^* = molecular diffusion coefficient in water

If adsorption on fracture surface is governed by linear isotherm,

$$s = \frac{ds}{dc} c = K_f c$$

25 D. H. Tang, E. O. Frind, and E. A. Sudicky (1981). Contaminant Transport in Fractured Porous Media: Analytical Solution for a Single Fracture. Water Resources Research, Vol. 17, pp. 555-564.

then

$$\frac{\partial s}{\partial t}=\frac{ds}{dc}\frac{\partial c}{\partial t}=K_f\frac{\partial c}{\partial t} \tag{3.14}$$

and the differential equation for the fracture ignoring diffusive loss term can be written as

$$\frac{\partial c}{\partial t}+\frac{v}{R}\frac{\partial c}{\partial z}-\frac{D}{R}\frac{\partial^2 c}{\partial z^2}+\lambda c+\frac{q}{BR}=0, \qquad 0\le z\le\infty \tag{3.15}$$

The mass balance equation for a strip of unit width, extending in the direction perpendicular to the fracture, yields the differential equation for the porous matrix:

$$-D'\frac{\partial^2 c}{\partial x^2}+\frac{\partial c'}{\partial t}+\frac{\rho_b}{\theta}\frac{\partial s'}{\partial t}+\lambda c'+\frac{\rho_b}{\theta}\lambda s'=0, \qquad b\le x<\infty \tag{3.16}$$

where

$x=$ coordinate perpendicualr to the fracture axis
$c'=c'(x,z,t)=$ solute concentration
$s'=s'(x,z,t)=$ solute mass adsorbed by unit mass of porous matrix
$\rho_b=$ bulk density of the matrix
$\theta=$ porosity
$D'=\tau D^*$ with τ is the matrix tortuosity

Assuming a linear isotherm for adsorption within the porous matrix, we get

$$s'=\frac{ds'}{dc'}c'=K_m c' \text{ which yields}$$
$$\frac{\partial s'}{\partial t}=\frac{ds'}{dc'}\frac{\partial c'}{\partial t}=K_m\frac{\partial c'}{\partial t} \tag{3.17}$$
where K_m is the mass distribution coefficient.

Equation (3.17) leads to the final differential equation for the matrix as

$$\frac{\partial c'}{\partial t}-\frac{D'}{R'}\frac{\partial^2 c'}{\partial x^2}+\lambda c'=0 \tag{3.18}$$

where the retardation coefficient is given by

$$R' = 1 + \frac{\rho_b}{\theta}.$$

The diffusive loss term is the mass flux crossing the fracture-matrix interface, which is governed by Fick's first law:

$$q = -\theta D' \left.\frac{\partial c'}{\partial x}\right|_{x=b} \tag{3.19}$$

Equations (3.15) and (3.19) yield the final equation for the fracture:

$$\frac{\partial c}{\partial t} + \frac{v}{R}\frac{\partial c}{\partial z} - \frac{D}{R}\frac{\partial^2 c}{\partial z^2} + \lambda c - \frac{\theta D'}{BR}\left.\frac{\partial c'}{\partial x}\right|_{x=b} = 0, \quad 0 \le z \le \infty \tag{3.20}$$

The boundary conditions for (3.18) and (3.20) are

$$\begin{aligned} c(0,t) &= c_0 \\ c(\infty,t) &= 0 \\ c(z,0) &= 0 \end{aligned} \tag{3.21}$$

wherec_0 is the source concentration, and

$$\begin{aligned} c'(b,z,t) &= c(z,t) \\ c'(\infty,z,t) &= 0 \\ c'(x,z,0) &= 0 \end{aligned} \tag{3.22}$$

General transient solution

The general transient solution of the coupled system of equations (3.18)under the boundary conditions (3.21) is obtained as follows.

Taking the Laplace transform of (3.18), we get

$$p\overline{c'} = \frac{D'}{R'}\frac{d^2\overline{c'}}{dx^2} - \lambda\overline{c'} \tag{3.23}$$

where

$$\overline{c'}(x,z,p)=\int_0^\infty \exp(-pt)c'(x,z,t)dt \tag{3.24}$$

is the Laplace transform of c'. The only admissible solution for (3.23) is

$$\overline{c'}=c_1'\exp\left\{-BP^{1/2}(x-b)\right\} \tag{3.25}$$

where

$B=(R'/D')^{1/2}$ and $P=p+\lambda$.

Substituting the value of the constant of integration c_1', obtained from the boundary conditions, we obtain

$$\overline{c'}=\overline{c}\exp\left\{-BP^{1/2}(x-b)\right\}. \tag{3.26}$$

The gradient of $\overline{c'}$ at the interface $x=b$ is

$$\left.\frac{d\overline{c'}}{dx}\right|_{x=b}=-BP^{1/2}\overline{c} \tag{3.27}$$

We next apply the Laplace transform to (3.20) and obtain

$$p\overline{c}+\frac{v}{R}\frac{d\overline{c}}{dz}+\lambda\overline{c}=\frac{\theta D'}{bR}\left.\frac{d\overline{c'}}{dx}\right|_{x=b}+\frac{D}{R}\frac{d^2\overline{c}}{dz^2} \tag{3.28}$$

From (3.27) and (3.28) we get

$$\frac{d^2\overline{c}}{dz^2}-\frac{v}{D}\frac{d\overline{c}}{dz}-\frac{R}{D}\left\{P+\frac{P^{1/2}}{A}\right\}\overline{c}=0 \tag{3.29}$$

where

$$A=\frac{bR}{\theta(R'D')^{1/2}}.$$

The second-order ordinary differential equation (3.29) has a solution of the form

$$\overline{c}=c_2\exp(zr_+)+c_3\exp(zr_-) \tag{3.30}$$

where c_2 and c_3 are undetermined constants and $r_\pm$ is given by

$$r_\pm = \nu\left[1 \pm \left\{1+\beta^2\left(\frac{P^{1/2}}{A}+P\right)\right\}^{1/2}\right] \tag{3.31}$$

where

$$\nu = \frac{V}{2D},\ \beta^2 = \frac{4RD}{V^2}.$$

For the solution to be finite, the first term in (3.30) must vanish, yielding

$$\bar{c} = c_3 \exp(\nu z)\exp\left[-\nu z\left\{1+\beta^2\left(\frac{P^{1/2}}{A}+P\right)\right\}^{1/2}\right] \tag{3.32}$$

Applying the Laplace transform to the boundary condition (3.21) we obtain

$$\bar{c}(0,p) = \frac{c_0}{p} = \frac{c_0}{P-\lambda} = c_{3.} \tag{3.33}$$

and hence

$$\bar{c} = \frac{c_0}{P-\lambda}\exp(\nu z)\exp\left[-\nu z\left\{1+\beta^2\left(\frac{P^{1/2}}{A}+P\right)\right\}^{1/2}\right] \tag{3.34}$$

To invert (3.34), we use the identity

$$\int_0^\infty \exp\left(-\xi^2-\frac{x^2}{\xi^2}\right)d\xi = \frac{\pi^{1/2}}{2}\exp(-2x)$$

and convert the exponential term

$$\exp\left[-\nu z\left\{1+\beta^2\left(\frac{P^{1/2}}{A}+P\right)\right\}^{1/2}\right]$$

into

$$\frac{2}{\pi^{1/2}}\int_0^\infty \exp\left[-\xi^2-\frac{\nu^2 z^2}{4\xi^2}\left\{1+\beta^2\left(\frac{P^{1/2}}{A}+P\right)\right\}\right]d\xi$$

and obtain

$$\frac{\overline{c}}{c_0}=\frac{2}{\pi^{1/2}}\exp(\nu z)\int_0^\infty \exp\left[-\xi^2-\frac{\nu^2 z^2}{4\xi^2}\right]\cdot\frac{\exp\left(-Y\left(P^{1/2}+AP\right)\right)}{P-\lambda}d\xi \tag{3.35}$$

where

$$Y=\frac{\nu^2\beta^2 z^2}{4A\xi^2}.$$

Following two results are needed to invert the Laplace transform:

$$\mathcal{L}^{-1}\left\{\frac{\exp\left(-p^{1/2}Y\right)}{p-\gamma}\right\}=\frac{1}{2}\exp(\gamma t)\{A_1+A_2\} \tag{3.36}$$

where

$$A_1=\exp\left[-\lambda^{1/2}Y\right]erfc\left[\frac{Y}{2t^{1/2}}-(\gamma t)^{1/2}\right]$$

$$A_2=\exp\left[\gamma^{1/2}Y\right]erfc\left[\frac{Y}{2t^{1/2}}+(\gamma t)^{1/2}\right]$$

erfc is the complementary error function,

and

$$\mathcal{L}^{-1}\{\exp(-PE)\cdot\varphi(P)\}=\varphi(t-E)U(t-E) \tag{3.37}$$

where

$$U=\begin{cases}0 & t<E\\ 1 & t\geq E.\end{cases}$$

Using the results (3.36) and (3.37), the inverse Laplace transform of (3.35) turns out to be

$$\frac{c}{c_0}=\frac{2}{\pi^{1/2}}\exp(\nu z)\int_0^\infty \exp\left[-\xi^2-\frac{\nu^2 z^2}{4\xi^2}\right]\frac{1}{2}U(T^2)\exp(-\eta z^2)$$
$$\times\left\{\exp\left[-\lambda^{1/2}Y\right]erfc\left[C_1\right]+\exp\left[\lambda^{1/2}Y\right]erfc(C_2)\right\}d\xi \tag{3.38}$$

where

$$C_1=\frac{Y}{2T}-\lambda^{1/2}T$$
$$C_2=\left[\frac{Y}{2T}+\lambda^{1/2}T\right].$$

Since $T\geq 0$, we have

$$\xi\geq\frac{z}{2}\left(\frac{R}{Dt}\right)^{1/2}.$$

and the final solution for concentration in the fracture is

$$\frac{c}{c_0}=\frac{\exp(\nu z)}{\pi^{1/2}}\int_\ell^\infty \exp\left(-\xi^2-\frac{\nu^2 z^2}{4\xi^2}\right)\exp(-\eta z^2)$$
$$\times\left\{\exp\left(-\lambda^{1/2}Y\right)\mathrm{erfc}(C_1)+\exp\left(\lambda^{1/2}Y\right)\mathrm{erfc}(C_2)\right\}d\xi, \tag{3.39}$$

wehere $\ell=\frac{z}{2}\left(\frac{R}{Dt}\right)^{1/2}.$

The inversion of $\overline{c}'$ yields the final solution for concentration in the porous matrix:

$$\frac{c'}{c_0}=\frac{\exp(\nu z)}{\pi^{1/2}}\int_\ell^\infty \exp\left[-\xi^2-\frac{\nu^2 z^2}{4\xi^2}\right]\cdot\exp(-\eta z^2)$$
$$\times\left\{\exp\left[-\lambda^{1/2}Y'\right]\cdot erfc\left[\frac{Y'}{2T}-\lambda^{1/2}T\right]+\exp\left[\lambda^{1/2}Y'\right]erfc\left[\frac{Y'}{2T}+\lambda^{1/2}T\right]\right\}d\xi \tag{3.40}$$

where

$$Y'=\frac{\nu^2 B^2 z^2}{4A\xi^2}+B(x-b).$$

Transient Solution with $D = 0$

The transient solution for the special case $D = 0$ cannot be obtained directly from the general solution, since (3.20) becomes singular as $D \to 0$. For this case, equation (3.28) becomes

$$p\bar{c} + \frac{v}{R}\frac{d\bar{c}}{dz} + \lambda\bar{c} = \frac{\theta D'}{bR}\frac{d\bar{c}'}{dx}\bigg|_{x=b} \tag{3.41}$$

Using (3.27), this becomes

$$\frac{v}{R}\frac{d\bar{c}}{dz} + \left\{p + \lambda + \frac{P^{1/2}}{A}\right\}\bar{c} = 0 \tag{3.42}$$

which has the solution

$$\bar{c} = \frac{c_0}{p}\exp\left(-\frac{\lambda Rz}{v}\right)\exp\left(-\frac{pRz}{v}\right)\exp\left(-\frac{RP^{1/2}z}{vA}\right) \tag{3.43}$$

The Laplace transform inversion yields the solution

$$\frac{c}{c_0} = \begin{cases} 0, & \text{if } T' < 0 \\ \frac{1}{2}\exp\left(-\frac{\lambda Rz}{v}\right)(W_1 + W_2), & \text{if } T' \geq 0 \end{cases} \tag{3.44}$$

where

$$W_1 = \exp\left(-\frac{\lambda^{1/2}Rz}{vA}\right)erfc\left(\frac{z}{2vAT} - \lambda^{1/2}T'\right)$$

$$W_2 = \exp\left(\frac{\lambda^{1/2}Rz}{vA}\right)erfc\left(\frac{z}{2vAT} + \lambda^{1/2}T'\right)$$

and

$$T' = \left(t - \frac{Rz}{v}\right)^{1/2}$$

The solution for the porous matrix is similarly obtained as

$$\frac{c}{c_0}=\begin{cases} 0, & \text{if } T'<0 \\ \frac{1}{2}\exp\left(-\frac{\lambda R z}{v}\right)\cdot\left[\exp\left(-\lambda^{1/2}W\right)erfc\left(\frac{W}{2T'}-\lambda^{1/2}T'\right)\right. & \\ \left.+\exp\left(\lambda^{1/2}W\right)erfc\left(\frac{W}{2T'}+\lambda^{1/2}T'\right)\right] & \text{if } T'\geq 0 \end{cases} \tag{3.45}$$

Example 3.3

The transient contaminant transport in a system of parallel fractures in a porous matrix is discussed in this example[26]. An exact solution for the following physical phenomena has been derived:

(i) Molecular diffusion and mechanical dispersion along the fracture axes,
(ii) Molecular diffusion from the fracture to the perpendicular porous matrix,
(iii) Adsorption onto the face of the matrix and within the matrix,
(iv) Radioactive decay,
(v) Transport through the rock matrix by advection is negligible.

The transport mechanism in the case of system of parallel fractures is described by the following coupled one-dimensional differential equations:

$$\frac{\partial c}{\partial t}+\frac{v}{R}\frac{\partial c}{\partial z}-\frac{D}{R}\frac{\partial^2 c}{\partial z^2}+\lambda c+\frac{q}{Rb}=0, \quad 0 \le z<\infty \tag{3.46}$$

$$\frac{\partial c'}{\partial t}-\frac{D'}{R'}\frac{\partial^2 c'}{\partial x^2}+\lambda c=0, \quad b\le x\le B \tag{3.47}$$

Let

$z=$ spatial coordinate along the fracture
$c'=c'(x,z,t)$ is the concentration in the porous matrix
$c=c(z,t)$ = concentration of solute in fracture
$2b=$ fracture width
$x=$ spatial coordinate perpendicular to the fracture axis
$2B=$ the fracture spacing
$\sigma=$ matrix porosity
$K_f=$ fracture distribution coefficient
$D*=$ molecular diffusion coefficient for solute in free solution
$D'=$ diffusion coefficient
$R'=$ matrix retardation coefficient = $1+\frac{P_b}{\sigma}K_m$
$K_m=$ porous matrix distribution coefficient

26 E. A. Sudicky and E. O. Frind (1982). Contaminant Transport in Fractured Porous Media: Analytical Solutions for a System of Parallel Fractures. Water Resources Research, Vol. 18, pp. 1634-1642.

$$q = -\theta D' \frac{\partial c'}{\partial x}\bigg|_{x=b} \quad \text{loss term}$$

$$R = \text{face retardation coefficient} = 1 + \frac{K_f}{b}$$

The final form of the equation describing transport in a fracture is

$$\frac{\partial c}{\partial t} + \frac{v}{R}\frac{\partial c}{\partial z} - \frac{D}{R}\frac{\partial^2 c}{\partial z^2} + \lambda c - \frac{\theta D'}{bR}\frac{\partial c'}{\partial x}\bigg|_{x=b} = 0, \quad 0 \le z < \infty \tag{3.48}$$

The initial and boundary conditions are

$$\begin{aligned} c(z,0) &= 0 \\ c(0,t) &= c_0 \\ c(\infty,t) &= 0 \end{aligned} \tag{3.49}$$

$$\begin{aligned} c'(x,z,0) &= 0 \\ c'(b,z,t) &= c(z,t) \\ \frac{\partial c'}{\partial x}(B,z,t) &= 0 \end{aligned} \tag{3.50}$$

The coupling of the matrix to the fracture is expressed in (3.50).

General Transient Solution

The Laplace transform of (3.47) is

$$p\overline{c'} - \frac{D'}{R'}\frac{d^2\overline{c'}}{dx^2}\overline{c'} + \lambda\overline{c'} = 0 \tag{3.51}$$

where $\overline{c'}$ is the Laplace transform of c'.
The general solution to the above equation has the form

$$\overline{c'} = c'_1 \cosh\left\{GP^{1/2}(B-x)\right\} + c'_2 \sinh\left\{GP^{1/2}(B-x)\right\} \tag{3.52}$$

where $G = (R'/D')^{1/2}$ and $P = p + \lambda$.

Boundary conditions (3.50) imply $c_2' = 0$, and the above solution reduces to

$$\bar{c}' = \bar{c}\frac{\cosh\{GP^{1/2}(B-x)\}}{\cosh\{\sigma P^{1/2}\}} \tag{3.53}$$

where $\sigma = G(B-b)$.

The gradient of $\bar{c}'$ at $x=b$ is

$$\left.\frac{d\bar{c}'}{dx}\right|_{x=b} = -\bar{c}GP^{1/2}\tanh\left(\sigma P^{1/2}\right) \tag{3.54}$$

Application of the Laplace transform to the transport equation for the fracture, and substitution of the interface gradient yields

$$\frac{d^2\bar{c}}{dz^2} - \frac{v}{D}\frac{d\bar{c}}{dz} - \frac{R}{D}\left\{P + \frac{P^{1/2}}{A}\tanh\left(\sigma P^{1/2}\right)\right\}\bar{c} = 0 \tag{3.55}$$

where $A = \dfrac{bR}{\theta\left(R'D'\right)^{1/2}}$.

The general solution of the above second order differential equation is

$$\bar{c} = c_1\exp\left(zr_+\right) + c_2\exp\left(zr_-\right) \tag{3.56}$$

where

$$r_\pm = \nu\left[1 \pm \left\{1 + k^2\left(\frac{P^{1/2}}{A}\tanh\left(\sigma P^{1/2}\right)\right)\right\}^{1/2}\right] \text{ and } \nu = \frac{v}{2D} \text{ and } k^2 = \frac{4RD}{v^2}.$$

Since the solution is bounded, c_1 in (3.56) must be 0;taking the Laplace transform of the boundary condition (3.49) we obtain

$$\bar{c}(0,p) = \frac{c_0}{p} = \frac{c_0}{P-\lambda} = c_2. \tag{3.57}$$

Substituting the value of c_2 in (3.56) leads to

$$\overline{c} = \frac{c_0}{P-\lambda}\exp(\nu z)\exp\left\{-\nu z\left[1+k^2\left(\frac{P^{1/2}}{A}\tanh(\sigma P^{1/2})+P\right)\right]^{1/2}\right\} \tag{3.58}$$

which simplifies to

$$\begin{aligned}\frac{\overline{c}}{c_0} &= \frac{2}{\pi^{1/2}}\exp(\nu z)\int_0^\infty \exp\left[-\xi^2-\frac{\nu^2 z^2}{4\xi^2}\right]\frac{1}{P-\lambda} \\ &\cdot\exp(-YAP)\exp\left[-YP^{1/2}\tanh\left(\sigma p^{1/2}\right)\right]d\xi\end{aligned} \tag{3.59}$$

where $Y = \dfrac{\nu^2 k^2 z^2}{4A\xi^2}$.

The concentration c in a fracture is the inverse Laplace transform of (3.59):

$$\begin{aligned}\frac{c}{c_0} &= \frac{2}{\pi^{1/2}}\exp(\nu z)\int_0^\infty \exp\left[-\xi^2-\frac{\nu^2 z^2}{4\xi^2}-\frac{R\lambda z^2}{4D\xi^2}\right]d\xi \\ &\cdot\int_0^T \mathcal{L}^{-1}\left[\exp\left\{-Yp^{1/2}\tanh\left(\sigma p^{1/2}\right)\right\}\right]\Big|_\tau \exp(-\lambda\tau)d\tau\end{aligned} \tag{3.60}$$

where $T = t-\dfrac{Rz^2}{4D\xi^2},\ T\geq 0.$

Using the fact that

$$L^{-1}\left\{\exp\left(-Yp^{1/2}\right)\tanh\left(\sigma p^{1/2}\right)\right\} = \frac{1}{\pi}\int_0^\infty \varepsilon\exp(\varepsilon_R)\cos(\varepsilon_\ell)d\xi \tag{3.61}$$

where

$$\varepsilon_R = -\frac{y_\varepsilon}{2}\left(\frac{\sinh(\sigma\varepsilon)-\sin(\sigma\varepsilon)}{\cosh(\sigma\varepsilon)+\cos(\sigma\varepsilon)}\right) \text{ and } \varepsilon_\ell = \frac{\varepsilon^2 t}{2}-\frac{Y_\varepsilon}{2}\left(\frac{\sinh(\sigma\varepsilon)+\sin(\sigma\varepsilon)}{\cosh(\sigma\varepsilon)+\cos(\sigma\varepsilon)}\right)$$

equation (3.60) becomes

$$\frac{c}{c_o} = \frac{2}{\pi^{1/2}} \exp(\nu z) \int_0^\infty \exp\left[-\xi^2 - \frac{\nu^2 z^2}{4\xi^2} - \frac{R\lambda z^2}{4D\xi^2}\right] \cdot \int_0^\infty \varepsilon \exp(\varepsilon_R) \cdot \int_0^T \exp(-\lambda\tau)\cos(\varepsilon_I) d\tau d\varepsilon d\xi \tag{3.62}$$

by interchanging the order of integration of τ and. Equation (3.62) can now be integrated with respect to τ to obtain

$$\frac{c}{c_o} = \frac{2}{\pi^{2/3}} \exp(\nu z) \int_0^\infty \exp\left[-\xi^2 - \frac{\nu^2 z^2}{4\xi^2} - \frac{R\lambda z^2}{4D\xi^2}\right] \cdot \int_0^\infty \frac{\varepsilon}{\lambda^2 + \varepsilon^4/4} \exp(\varepsilon_R) \left[\exp(-\lambda T)\left\{\frac{\varepsilon^2}{2}\sin(\varepsilon_I)|\tau - \lambda\cos(\varepsilon_I)|_T\right\} + \frac{\varepsilon^2}{2}\sin(\Omega) + \lambda\cos(\Omega)\right] d\varepsilon d\xi \tag{3.63}$$

where

$$\Omega = \frac{Y_\varepsilon}{2}\left(\frac{\sinh(\sigma\varepsilon) + \sin(\sigma\varepsilon)}{\cosh(\sigma\varepsilon) + \cos(\sigma\varepsilon)}\right) \text{ and } \xi \geq \frac{z}{2}\left(\frac{R}{Dt}\right)^{1/2}.$$

The final form of the solution is

$$\frac{c}{c_o} = \frac{2}{\pi^{2/3}} \exp(\nu z) \int_\ell^\infty \exp\left[-\xi^2 - \frac{\nu^2 z^2}{4\xi^2} - \frac{R\lambda z^2}{4D\xi^2}\right] \cdot \int_0^\infty \frac{\varepsilon}{\lambda^2 + \varepsilon^4/4} \exp(\varepsilon_R) \left[\exp(-\lambda T)\left\{\frac{\varepsilon^2}{2}\sin(\varepsilon_I)\Big|_T - \lambda\cos(\varepsilon_I)\Big|_T\right\} + \frac{\varepsilon^2}{2}\sin(\Omega) + \lambda\cos(\Omega)\right] d\varepsilon d\xi \tag{3.64}$$

where $\ell = \frac{z}{2}\left(\frac{R}{Dt}\right)^{1/2}$ is the lower limit of integration.

Next, we find the solution for the concentration in the porous matrix. Using the fact that

$$\mathcal{L}^{-1}\left[\frac{\cosh\left(G(B-x)^{1/2}\right)}{\cosh\left(\sigma P^{1/2}\right)}\right]=\frac{\pi}{\sigma^2}\sum_{n=0}^{\infty}(-1)^n(2n+1)\cdot\exp\left[-\left(\frac{\pi^2(2n+1)^2}{4\sigma^2}+\lambda\right)t\right]\cdot\cos\left[\frac{(2n+1)\pi(B-x)}{2(B-b)}\right] \tag{3.65}$$

the inverse Laplace transform of (3.53) can be written as

$$\frac{c'}{c}=\frac{\pi}{\sigma^2}\sum_{n=0}^{\infty}(-1)^n(2n+1)\cos\left[\frac{(2n+1)\pi(B-x)}{2(B-b)}\right]\cdot\int_0^t c(z,\tau)\exp\left[-\left(\frac{\pi^2(2n+1)^2}{4\sigma^2}+\lambda\right)(t-\tau)\right]d\tau \tag{3.66}$$

Substitution of (3.64) into (3.66) for c yields

$$\begin{aligned}\frac{c'}{c_0}=&\frac{2}{\pi^{1/2}\sigma^2}\exp(\nu z)\int_{\ell}^{\infty}\exp\left[-\xi^2-\frac{\nu^2z^2}{4\xi^2}-\frac{R\lambda z^2}{4D\xi^2}\right]\\&\cdot\int_0^{\infty}\frac{\varepsilon}{\lambda^2+\varepsilon^4/4}\exp(\varepsilon_R)\sum_{n=0}^{\infty}(-1)^n(2n+1)\cos\left[\frac{(2n+1)\pi(B-x)}{2(B-b)}\right]\\&\cdot\left[\frac{\exp(-\lambda T)}{\pi^4(2n+1)^4/16\sigma^2+\varepsilon^4/4}\left\{\frac{\varepsilon^2}{2}\left(\frac{\pi^2(2n+1)^2}{4\sigma^2}\sin(\varepsilon_I)\Big|_T-\frac{\varepsilon^2}{2}\cos\varepsilon_I\Big|_T\right)\right.\right.\\&-\lambda\left(\frac{\pi^2(2n+1)^2}{4\sigma^2}\cos(\varepsilon_I)\Big|_T+\frac{\varepsilon^2}{2}\sin(\varepsilon_I)\Big|_T\right)\\&+\exp\left[-\frac{\pi^2(2n+1)^2t}{4\sigma^2}\right]\cdot\left(\frac{\varepsilon^4}{4}\cos(\Omega')+\frac{\pi^2(2n+1)^2}{8\sigma^2}\varepsilon^2\sin(\Omega')\right.\\&\left.\left.+\frac{\lambda\pi^2(2n+1)^2}{4\sigma^2}\cos(\Omega')-\frac{\lambda\varepsilon^2}{2}\sin(\Omega')\right)\right\}\\&\left.+\frac{4\sigma^2}{\pi^2(2n+1)^2}\left\{1-\exp\left[-\frac{\pi^2(2n+1)^2t}{4\sigma^2}\right]\right\}\left\{\frac{\varepsilon^2}{2}\sin(\Omega)+\lambda\cos(\Omega)\right\}\right]d\varepsilon d\xi\end{aligned} \tag{3.67}$$

where $\Omega'=\Omega+\dfrac{Rz^2\varepsilon^2}{8\xi^2}$.

Transient Solution with D=0

Equations (3.47) and (3.50) for the porous matrix remain unchanged for this case. The Laplace transform of the equation (3.48) becomes

$$\frac{d\bar{c}}{dz}+\left[\frac{RP}{\nu}+\frac{\theta\left(R'D'P\right)^{1/2}}{b\nu}\tanh\left(\sigma P^{1/2}\right)\right]\bar{c}=0. \tag{3.68}$$

The solution in this case subject to (3.49) is

$$\frac{c'}{c_0}=\frac{c_0}{P-\lambda}\exp\left(-\frac{RPz}{\nu}\right)\exp\left(-\omega P^{1/2}\tanh\left(\sigma P^{1/2}\right)\right) \tag{3.69}$$

where

$$w=\frac{\theta\left(R'D'\right)^{1/2}z}{b\nu}.$$

This is inverted in the same manner as in the previous case, yielding the concentration in a fracture as

$$\frac{c}{c_0}=\begin{cases}0, & T^0<0\\ \frac{1}{\pi}\int_0^\infty H_1(\varepsilon)\exp\left(\varepsilon_R^0\right)H_2(\varepsilon)\,d\varepsilon, & T^0>0\end{cases} \tag{3.70}$$

where

$$H_1(\varepsilon)=\exp\left(\frac{R\lambda z}{\nu}\right)\frac{\varepsilon}{\lambda^2+\varepsilon^4/4}$$

$$H_2(\varepsilon)=\exp\left(-\lambda T^0\right)\left(\frac{\varepsilon^2}{2}\sin\left(\varepsilon_I^0\right)-\lambda\cos\left(\varepsilon_I^0\right)\right)+\frac{\varepsilon^2}{2}\sin\left(\Omega^0\right)+\lambda\cos\left(\Omega^0\right)$$

$$T^0=t-\frac{Rz}{\nu},\ \varepsilon_R^0=-\frac{\omega\varepsilon}{2}\left(\frac{\sinh(\sigma\varepsilon)-\sin(\sigma\varepsilon)}{\cosh(\sigma\varepsilon)+\cos(\sigma\varepsilon)}\right)$$

$$\varepsilon_I^0=\frac{\varepsilon^2T^0}{2}-\Omega^0,\ \Omega^0=\frac{\omega\varepsilon}{2}\left(\frac{\sinh(\sigma\varepsilon)+\sin(\sigma\varepsilon)}{\cosh(\sigma\varepsilon)+\cos(\sigma\varepsilon)}\right).$$

The solution for the porous matrix can similarly be obtained by using equation (3.67).

$$\frac{c'}{c_0}=\begin{cases}0, & T^0<0\\ \frac{1}{\sigma^2}\int_0^\infty H_1(\varepsilon)\exp\left(\varepsilon_R^0\right)F_1\left[F_2\left\{F_3+F_4+F_5\right\}+F_6\right]d\varepsilon, & T^0>0\end{cases} \tag{3.71}$$

where

$$F_1=\sum_{n=0}^{\infty}(-1)^n(2n+1)\cos\left[\frac{(2n+1)\pi(B-x)}{2(B-b)}\right]$$

$$F_2=\frac{\exp\left(-\lambda T^0\right)}{\pi^4(2n+1)^4/16\sigma^4+\varepsilon^4/4}$$

and

$$F_3=\frac{\varepsilon^2}{2}\left(A\sin\left(\varepsilon_I^0\right)-\frac{\varepsilon^2}{2}\cos\left(\varepsilon_I^0\right)\right)$$

$$F_4=-\lambda\left(A\cos\left(\varepsilon_I^0\right)+\frac{\varepsilon^2}{2}\sin\left(\varepsilon_I^0\right)\right)+\exp(-At)\left(\frac{\varepsilon^4}{4}\cos\left(\Omega_0'\right)\right)$$

$$F_5=\frac{A}{2}\sin\left(\Omega_0'\right)+\lambda A\cos\left(\Omega_0'\right)-\frac{\lambda\varepsilon^2}{2}\sin\left(\Omega_0'\right)$$

$$F_6=\frac{1}{A}\{1-\exp[-At]\}\left\{\frac{\varepsilon^2}{2}\sin(\Omega_0)+\lambda\cos(\Omega_0)\right\}\Bigg]$$

$$A=\frac{\pi^2(2n+1)^2}{4\sigma^2}\text{ and }\Omega_0'=\Omega_0+\frac{Rz\varepsilon^2}{2v}.$$

Example 3.4

In this example[27], the problem of radionuclide migration in fissured rock is considered when the transport mechanism is advection and longitudinal dispersion in the fissures, external and internal diffusion into rock (assumed to be spherical blocks), sorption into the fissure surfaces, sorption within the matrix, and radioactive decay.

The mathematical model for flow and sorption from water in the fissures is

$$R_a \frac{\partial C_f}{\partial t} + U_f \frac{\partial C_f}{\partial z} - D_L \frac{\partial^2 C_f}{\partial z^2} = -\frac{3N_0}{mr_0} - R_a \lambda_d C_f \tag{3.72}$$

And that for diffusion and sorption in spherical rock blocks is

$$K\frac{\partial C_p}{\partial t} = D_p \varepsilon_p \left(\frac{\partial^2 C_p}{\partial r^2} + \frac{2}{r}\frac{\partial C_p}{\partial r} \right) - K\lambda_d C_p \tag{3.73}$$

where

$R_a = 1 + 3K_f / mr_0 =$ external sufrace sorption coefficient ($R_a = 1$ for nonabsorbing species)

$K_f =$ surface sorption coefficient

$\frac{3}{mr_0} =$ interfacial rock surface per unit fracture volume

$D_p =$ diffusivity in water in micropores $= D_v \frac{\delta_D}{\tau^2}$

$D_v =$ diffusivity in pure water

$\frac{\delta_D}{\tau^2} < 1$ is a geometric factor

$K =$ the volume equilibrium constant $= \varepsilon_p + K_a$.

In general, the exterior and interior surfaces are governed by different linear equilibrium relationships.

27 Anders Rasmuson (1984). Migration of Radionuclides in Fissured Rock: Analytical Solutions for the Case of Constant Source Strength. Water Resources Research, Vol. 20, pp. 1435-1442.

If a system is initially free of nuclides, and has inlet nuclide concentration C_0 at time zero and kept constant, then the initial and boundary conditions are

$$\begin{aligned} &C_f(0,t)=C_o \\ &C_f(\infty,t)=0 \\ &C_f(z,0)=0 \\ &\left(\frac{\partial C_p}{\partial r}\right)(0,z,t)=0 \\ &N_o=D_p\varepsilon_p\left(\frac{\partial C_p}{\partial_r}\right)_{r=r_o}=K_f\left(C_f-C_p|_{r=r_o}\right) \\ &C_p(r,z,0)=0 \end{aligned} \tag{3.74}$$

where N_0 is the mass flux from the flowing phase to the outer block surfaces. When film diffusion is negligible $(k_f\rightarrow\infty)$,

$$C_p(r_o,z,t)=C_f(z,t) \tag{3.75}$$

Appling the Laplace transform to (3.74), we obtain

$$Ks\bar{C}_p=D_p\varepsilon_p\left(\frac{d^2\bar{C}_p}{dr^2}+\frac{2}{r}\frac{d\bar{C}_p}{dr}\right)-K\lambda_d\bar{C}_p. \tag{3.76}$$

Under the boundary conditions (3.75), the solution is:

$$\bar{C}_p=\left[r_o\sinh(Ar)\right]\left[r\sinh(Ar_o)\right]^{-1}\left(\bar{C}_p|_{r=r_o}\right) \tag{3.77}$$

where

$$A=\left(\frac{s+\lambda_d}{D_a}\right)^{1/2} \text{ and } D_a=D_p\varepsilon_p|K.$$

The derivative of (3.78) with respect to r is

$$\left.\frac{d\bar{C}_p}{dr}\right|_{r=r_o}=\frac{r_o^2A\coth(Ar_o)-r_o}{r_o^2}\left(\bar{C}_p\big|_{r=r_o}\right)=\frac{r_o}{3D_p\varepsilon_p}Y_D(s)\left(\bar{C}_p\big|_{r=r_o}\right) \tag{3.78}$$

where

$$Y_D(s)=\gamma\left[w(s)\coth w(s)-1\right]$$

$$w(s)=\left[v(s)\right]^{1/2},\ v(s)=r_0^2\frac{s+\lambda_d}{D_a},\ \gamma=\frac{3D_p\varepsilon_p}{r_0^2}.$$

Substituting (3.79)in the boundary condition (3.75) we get

$$\left.\bar{C}_p\right|_{r=r_0}=\frac{\bar{C}_f}{R_F y_D+1}$$
$$\bar{N}_0=\frac{r_0}{3}Y_T(s)\bar{C}_f \tag{3.79}$$

where

$$Y_T(s)=\frac{Y_D(s)}{R_F Y_D(s)+1},\quad R_f=\frac{r_0}{3k_f}.$$

In case film resistance is negligible,

$$\left.\bar{C}_p\right|_{r=r_0}=\bar{C}_f \text{ and } Y_T(s)=Y_D(s).$$

Taking the Laplace transform of (3.73) and using (3.80) we get

$$\frac{d^2\bar{C}_f}{dz^2}-\frac{U_f}{D_L}\frac{d\bar{C}_f}{dz}-\left(\frac{R_a(s+\lambda_d)}{D_L}+\frac{Y_T(s)}{mD_L}\right)\bar{C}_f=0. \tag{3.80}$$

Solving (3.81) with the boundary conditions to (3.75), we obtain

$$\bar{u}=\frac{\bar{C}_f}{C_0}=\frac{1}{s}\exp\left\{\left[\frac{U_f}{2D_L}-\left(\frac{U_f^2}{4D_L^2}+\frac{R_a(s+\lambda_d)}{D_L}+\frac{Y_T(s)}{mD_L}\right)^{1/2}\right]z\right\} \tag{3.81}$$

The application of the complex inversion formula of Laplace transformation yields

$$u=\frac{C_f}{C_0}=\exp\left(\frac{U_f z}{2D_L}\right)\frac{1}{2\pi i}\int_{a-i\infty}^{a+i\infty}\frac{1}{s}\exp\left[st-z\left(\frac{u_f^2}{4D_L^2}+\frac{R_a(s+\lambda d)}{D_1}+\frac{Y_T(s)}{mD_L}\right)^{1/2}\right]ds \tag{3.82}$$

Next, we need

$$\begin{aligned}
v(0) &= \lim_{|s|\to 0} v(s) = \frac{Kr_0^2\lambda_d}{\dot{D}_p\varepsilon_p} \\
Y_D(0) &= \lim_{|s|\to 0} Y_D(s) = \gamma\left\{\left[v(0)\right]^{1/2}\coth\left[v(0)\right]^{1/2} - 1\right\} \\
Y_T(0) &= \lim_{|s|\to 0} Y_T(s) = \frac{Y_D(0)}{R_F Y_D(0)+1} = \frac{1}{R_F}X_T(0) \\
X_T(0) &= 1 - \frac{1}{R_F Y_D(0)+1}.
\end{aligned} \tag{3.83}$$

By using polar coordinates, we can show that the integral along the imaginary axis, with small semi-circle Γ of radiusεabout the origin approaches

$$\frac{1}{2}\exp\left\{\left[\frac{U_f}{2D_L} - \left(\frac{U_f^2}{4D_L^2} + \frac{R_a\lambda_d}{D_L} + \frac{X_T(0)}{mD_L R_F}\right)^{1/2}\right]z\right\} \tag{3.84}$$

as $\varepsilon \to 0$.

The transform $\bar{C}_f$ is inverted to obtain:

$$u(z,t) = \frac{C_f(z,t)}{C_0} = \frac{1}{2}u_\infty + \frac{2}{\pi}\int_0^\infty \exp\left(\frac{U_f z}{2D_L} - A\right)\sin\left(\frac{2D_p\varepsilon_p}{Kr_0^2}\lambda^2 t - A\right)\frac{d\lambda}{\lambda} \tag{3.85}$$

where

$$A = z\left[\frac{\left(x'(\lambda)^2 + y'(\lambda)^2\right)^{1/2} + x'(\lambda)}{2}\right]^{1/2}$$

$$u_\infty = \exp\left\{\frac{u_f z}{2D_L} - \left[\frac{u_f^2 z^2}{4D_L^2} + \frac{R_a\lambda_d z^2}{D_L} + \frac{X_T(0)z^2}{mD_L R_F}\right]^{1/2}\right\}$$

$$X_T(0) = 1 - \left\{R_F\left[v(0)^{1/2}\coth\left(v(0)\right)^{1/2} - 1\right] + 1\right\}^{-1}$$

$$v(0) = Kr_0^2\lambda_d / D_p\varepsilon_p$$

$$z^2x' = \frac{u_f^2z^2}{4D_L^2} + \frac{vz^2}{mD_L}H_1 + \frac{R_a\lambda dz^2}{D_L}$$

$$z^2y' = \frac{2D_p\varepsilon_pR_az^2}{Kr_o^2D_L}\lambda^2 + \frac{vz^2}{mD_L}H_2$$

$$H_1 = \frac{H_{D_1} + v\left(H_{D_1}^2 + H_{D_2}^2\right)}{\left(1+vH_{D_1}\right)^2 + \left(vH_{D_2}\right)^2},\ H_2 = \frac{H_{D_2}}{\left(1+vH_{D_1}\right)^2 + \left(vH_{D_2}\right)^2}$$

$$H_{D_1} = \frac{\phi_2 \sinh 2\phi_2 + \phi_1 \sin 2\phi_1}{\cosh 2\phi_2 - \cos 2\phi_1},\ H_{D_2} = -\frac{\phi_1 \sin 2\phi_2 + \phi_2 \sin 2\phi_1}{\cosh 2\phi_2 - \cos 2\phi_1}$$

$$\phi_1 = -\left(\frac{\ell - c}{2}\right)^{1/2},\ \phi_2 = +\left(\frac{\ell + c}{2}\right)^{1/2}$$

$$\ell = \left(c^2 + d^2\right)^{1/2},\ c = Kr_o^2\lambda_o^2 / D_p\varepsilon_p,\ d = 2\lambda^2.$$

Example 3.5

In this example[28], the migration of radionuclides in the fissures in the bedrock surrounding a nuclear waste repository is considered. A one-dimensional transport model that includes radionuclide diffusion into the microfissures of the rock, linear sorption, and longitudinal dispersion in the bedrock is presented.

In mathematical development of the model, following assumptions are made: (1) The block diameter is small relative to the overall distance; (2) bedrock is macroscopically uniform; (3) the sorption equilibrium is linear, i.e., radionuclides are present in trace amounts, and local sorption equilibrium is reached instantaneously.

The equation for flow and sorption from the water in the fissures accounting for accumulation in the blocks, and radioactive decay is given by

$$\frac{\partial C_f}{\partial t} + U_f \frac{\partial C_f}{\partial z} - D_L \frac{\partial^2 C_f}{\partial z^2} = -\frac{1-\epsilon_f}{\epsilon_f}\left(\frac{\partial \overset{\Delta}{q}}{\partial t}\right) - \lambda_d C_f \qquad 3.86)$$

Fick's law of diffusion for radial movement of solute in a spherical particle (including radioactive decay) is

$$\frac{\partial q_i}{\partial t} = D_a\left(\frac{\partial^2 q_i}{\partial r^2} + \frac{2}{r}\frac{\partial q_i}{\partial r}\right) - \lambda_a q_i \qquad (3.87)$$

In this case, the local solute concentration term q_i, includes solute both in the solid rock and in the water in the pores. Assuming that the driving force for diffusion is the intrapore concentration gradient $\frac{\partial C_p}{\partial_r}$, an alternate differential equation for intraparticle diffusion is

$$\varepsilon_p \frac{\partial C_D}{\partial t} + \frac{\partial C_s}{\partial t} = \varepsilon_b D_p\left(\frac{\partial^2 C_p}{\partial r^2} + \frac{2}{r}\frac{\partial C_p}{\partial r}\right) - \lambda_d\left(\varepsilon_p C_p + C_s\right) \qquad (3.88)$$

28 Rasmuson and Neretnieks . Migration of Radionuclides in Fissured Rock, pp. 3749-3758.

The terms on the left-hand side give the accumulation in the pore fluid phase and in the solid phase, respectively, and the terms on the right – hand side describe diffusion in the pre fluid phase, and radioactive decay of the nuclide in the pores and solid rock.

By definition,

$$q_i = \varepsilon_p C_p + C_s$$

and from assumption 3

$$q_i = KC_p.$$

Therefore,

$$C_s = \left(K - \varepsilon_p\right) C_p \tag{3.89}$$

Differentiating (3.90) and substituting into (3.89), we obtain

$$\frac{\partial C_p}{\partial t} = \frac{\varepsilon_p D_p}{k}\left(\frac{\partial^2 C_p}{\partial r^2} + \frac{2}{r}\frac{\partial C_p}{\partial r}\right) - \lambda_d C_p. \tag{3.90}$$

If a system is initially free of nuclides, and has inlet nuclide concentration C_0 at time zero and kept constant, then the initial and boundary conditions are

$$\begin{aligned} &C_f\left(0,t\right) = C_0 e^{-\lambda_d t} \\ &C_f\left(\infty,t\right) = 0 \\ &C_f\left(z,0\right) = 0 \\ &C_p\left(0,z,t\right) < \infty \\ &C_p\left(r,z,0\right) = C_p\left(r,z,0\right) = 0 \end{aligned} \tag{3.91}$$

$$C_p\left(r_0,z,t\right) = C_p\big|_{r=r_0} \text{ given by } \frac{\partial \overset{\Delta}{q}}{\partial t} = \frac{3k_f}{r_0}\left(C_f - C_p\big|_{r=r_0}\right). \tag{3.92}$$

Boundary condition (3.93) states that the mass entering or leaving the blocks must equal the flow of mass transported across a stagnant fluid film at the external surface and provides the link between (3.87) and (3.91). Boundary condition (3.92) is used here because it describes a constant leach rate for a body containing a decaying nuclide, and is an important case to study.

An exact solution of (3.87) and (3.91 subject to the boundary conditions (3.92) and (3.93) is obtained in the following steps: (1) The solution for a stable species ($\lambda_d = 0$) is obtained; (2) The results for a finite release time are obtained using the Heaviside function; (3) The radionuclide delay is taken into consideration. The solution for a stable species is

$$\frac{C_f}{C_0} = \frac{1}{2} + \frac{2}{\pi}\int_0^\infty \frac{1}{\lambda}\exp\left(\frac{U_f z}{2D_L} - A\right)\sin\left\{\sigma\lambda^2 t - A\right\} d\lambda \tag{3.93}$$

where

$$A = z\sqrt{\frac{\sqrt{\left(x'(\lambda)^2 + y'(\lambda)^2\right)^{1/2}} + x'(\lambda)}{2}}$$

$$x'(\lambda) = \frac{U_f^2}{4D_L^2} + \frac{\gamma}{mD_L}H_1,\ y'(\lambda) = \frac{\sigma\lambda^2}{D_L} + \frac{\gamma}{mD_L}H_2$$

$$m = \frac{\in_f}{1-\in_f},\ \gamma = \frac{3D_a k}{r_0^2},\ \sigma = \frac{2D_a}{r_0^2},\ R_F = \frac{r_0}{3k_f}.$$

andH_1, H_2 are defined as functions of λ and $\nu = \gamma R_F$:

$$\begin{aligned} H_1(\lambda,\nu) &= \frac{H_{D_1} + \nu\left(H_{D_1}^2 + H_{D_2}^2\right)}{\left(1+\nu H_{D_1}\right)^2 + \left(\nu H_{D_2}\right)^2} \\ H_2(\lambda,\nu) &= \frac{H_{D_2}}{\left(1+\nu H_{D_1}\right)^2 + \left(\nu H_{D_2}\right)^2} \end{aligned} \tag{3.94}$$

$$\begin{aligned} H_{D_1}(\lambda) &= \lambda\left(\frac{\sinh 2\lambda + \sin 2\lambda}{\cos 2\lambda - \cos 2\lambda}\right) - 1 \\ H_{D_2}(\lambda) &= \lambda\left(\frac{\sinh 2\lambda - \sin 2\lambda}{\cosh 2\lambda - \cos 2\lambda}\right). \end{aligned} \tag{3.95}$$

For small values ofλ ($\lambda < 0.1$), the following approximations have relative error < 0.001:

$$H_{D_1} = 4\lambda^4 / 45$$
$$H_{D_2} = 2\lambda^2 / 3. \tag{3.96}$$

It follows that

$\lim_{\lambda \to 0} H_{D_1} = 0$ and $\lim_{\lambda \to 0} H_{D_2} = 0.$

Moreover, for high values of λ (λ> 5), the following approximations have relative error < 0.001:

$$H_{D_1} = \lambda - 1$$
$$H_{D_2} = \lambda. \tag{3.97}$$

It is interesting to note that in the case of a finite step boundary condition (implying that the inlet concentration is decreased to 0 again at timet_0)

$$C_f'(0,t) = \begin{cases} C_0 & 0 \le t \le t_0 \\ 0 & t > t_0 \end{cases} \tag{3.98}$$

The Laplace transform of (3.99) is

$$\overline{C_f'}(z,t) = \left(1 - e^{-t_0 s}\right)\overline{C}_f(z,s) \tag{3.99}$$

with the inverse given by

$$C_f'(z,t) = C_f(z,t) - C_f(z,t-t_0)H(t-t_0). \tag{3.100}$$

Example 3.6

Chen[29] (1986) derived analytical solutions for two models: Model 1, in which radionuclide material is assumed to be transported through the fracture by radial advection and longitudinal dispersion, and Model II, in which only radial advection is considered. Both of these models take into account nuclide leakage from fracture to porous matrix by one–dimensional molecular diffusion. Analytical solutions for the models (transient and steady state) are obtained subject to two different boundary conditions for the first model; analytical solutions for steady state condition and assumption solution for small-time periods are obtained; for intermediate and large time periods, numerical inversion of Laplace transform is used. Assumptions for Model 1 are:

1. Fracture is horizontal with infinite areal extent.
2. Injection well has radius r_0 and injection rate Q is constant.
3. Transport of contaminants through the fracture is by radial convection and longitudinal dispersion, and into the porous matrix is by molecular diffusion.
4. The velocity of the ground water flow due to injection is $V = A/r$.
5. Transverse dispersion effect is negligible.

Under these assumptions, the governing equation for concentration distributions in the porous matrix and fracture are:

$$D_{m2}\frac{\partial^2 C_2}{\partial z^2} - \lambda C_2 - \frac{\rho_b}{n}\frac{\partial S_2}{\partial t} - \frac{\rho_b}{b}\lambda C_2 = \frac{\partial C_2}{\partial t} \tag{3.101}$$

$$\frac{1}{r}\frac{\partial}{\partial r}\left(rD_r\frac{\partial C_1}{\partial r}\right) - V\frac{\partial C_1}{\partial r} - \frac{\dot{Q}}{b} - \frac{1}{b}\frac{\partial S_1}{\partial t} - \lambda S_1 - \frac{\lambda S_1}{b} = \frac{\partial C_1}{\partial t} \tag{3.102}$$

After taking into consideration the successive decay reactions underlying radionuclide transport, the equations (3.102) and (3.103) can be rewritten as

$$D_{m2}\frac{\partial^2 C_2}{\partial z^2} - \lambda R_2 C_2 = R_2\frac{\partial C_2}{\partial t} \tag{3.103}$$

---

29 Chia-Shyun Chen (1986). Solutions for Radionuclide Transport from an Injection Well into a Single Fracture in a Porous Formation. Water Resources Research, Vol. 22, pp. 508-518.

$$\frac{dA}{r}\frac{\partial^2 C_1}{\partial r^2}-\frac{A}{r}\frac{\partial C_1}{\partial r}+\frac{n_2 D_{m2}}{b}\frac{\partial C_2}{\partial z}\bigg|_{z=0}-\lambda R_1 C_1=R_1\frac{\partial C_1}{\partial t} \tag{3.104}$$

where

$\lambda =$ decay constant

$\rho_b =$ bulk density of the matrix

$D_r = dV = dA/r$

$d =$ fracture dispersivity

$A=\dfrac{Q}{4\pi b}=$ advection parameter

$S_1 = K_1 C_1$

$S_2 = K_2 C_2$

$\dot{Q}=-\dfrac{n_2 D_{m2}}{b}\dfrac{\partial C_2}{\partial z}\bigg|_{z=0}$

$R_i =$ retardation factors for the absorption reaction ($i=1$ for fracture, $i=2$ fpr [porous matrix

$R_1 = 1 + K_1/b$

$R_2 = 1 + \rho_b K_2/n_2$

$b =$ half fracture aperture

$K_1 =$ distribution coefficient (mass of solute absorbed by unit area of solid surface divided by the concentration of solute in solution)

$K_2 =$ similarly defined for porous matrix

The initial conditions are:

$$C_1(r,0)=C_2(z,0)=0.$$

The boundary conditions at the interface of the fracture and the porous matrix are given as the condition of continuity of concentration,

$$C_1(r,t)=C_2(0,t).$$

The boundary conditions at r = z =∞are

$$C_j(\infty,t)=0,\ j=1,2.$$

Two different boundary conditions are considered at the well bore:

$$C_1(r_0,t)=e^{-\lambda t}$$

which corresponds to the injection of a radioactive substance with a short half-life. The second boundary condition is:

$$C_1(r_0,t)=1.$$

which assumes that the concentration at the injection well remains constant.

Solution for decay boundary.
The governing equations in dimensionless from are:

$$\frac{\partial^2 C_2}{\partial \xi^2}-\alpha_1 C_2=\frac{\partial C_2}{\partial \tau} \tag{3.105}$$

$$\frac{1}{\rho}\frac{\partial^2 C_1}{\partial \rho^2}-\frac{1}{\rho}\frac{\partial C_1}{\partial \rho}+\alpha\left.\frac{\partial C_2}{\partial \xi}\right|_{\xi=0}-\alpha_1 C_1=\frac{\partial C_1}{\partial \tau} \tag{3.106}$$

and the initial and boundary conditions become

$$C_1(\rho,0)=C_2(\xi,0)=0$$

$$C_1(\rho,\tau)=C_2(0,\tau)$$

$$C_1(\infty,\tau)=C_2(\infty,\tau)=0$$

$$C_1(p_0,\tau)=\exp(-\alpha_1 t)$$

whereρ, ρ_0, ξ, and τ are the dimensionless radial distance, the well radius, the vertical distance, and the time, respectively;$\alpha_1$is related to the radioactive decay, andα is a dimensionless ratio of the diffusive loss effect to the injection effect.

Applying the Laplace transform with respect toτwe get

$$\frac{d^2 G_2}{d\xi^2}-(p+\alpha_1)G_2=0 \tag{3.107}$$

$$\frac{1}{\rho}\frac{d^2G_1}{d\rho^2}-\frac{1}{\rho}\frac{dG_1}{d\rho}+\alpha\left.\frac{dG_2}{d\xi}\right|_{\xi=0}-(p+\alpha_1)G_1=0 \tag{3.108}$$

with

$$G_1(\rho,p)=G_2(0,p)$$
$$G_1(\infty,p)=G_2(\infty,p)=0$$
$$G_1(\rho_0,p)=1/(p+\alpha_1)$$

$$G_1(\rho,p)=\int_0^\infty e^{-p\tau}C_1(p,\tau)d\tau$$
$$G_2(\xi,p)=\int_0^\infty e^{-p\tau}C_2(\xi,\tau)d\tau$$

The solution is given by

$$G_2(\xi,p)=G_1\exp\left[-\xi(p+\alpha_1)^{1/2}\right]. \tag{3.109}$$

The gradient of G_2 at $\xi = 0$ is

$$\frac{dG_2}{d\xi}=-(p+\alpha_1)^{1/2}G_1. \tag{3.110}$$

Substituting (3.111) in (3.109) yields

$$\frac{1}{\rho}\frac{d^2G_1}{d\rho^2}-\frac{1}{\rho}\frac{dG_1}{d\rho}-\left[p+\alpha_1+\alpha(p+\alpha_1)^{1/2}\right]G_1=0. \tag{3.111}$$

The solution in terms of Airy's function is

$$G_1(\rho,p)=\frac{1}{\rho+\alpha_1}\exp\left(\frac{\rho-\rho_0}{2}\right)\frac{Ai(\beta^{1/3}y)}{Ai(\beta^{1/3}y_0)}. \tag{3.112}$$

It follows from (3.111) and (3.113) that

$$G_2(\xi,p)=\frac{1}{p+\alpha_1}\exp\left(\frac{\rho-\rho_0}{2}\right)\exp\left[-\xi(p+\varepsilon_1)^{1/2}\right]\frac{Ai(\beta^{1/3}y)}{Ai(\beta^{1/3}y_0)}. \tag{3.113}$$

For small τ or large p, the Airy's function can be approximated by

$$Ai(x) \simeq \left[2(\pi)^{1/2}\right]^{-1} x^{-1/4} \exp\left(-\frac{2}{3}x^{3/2}\right), \quad x >> 1 \tag{3.114}$$

Using the shifting rule of Laplace inversion, we obtain

$$C_1 = F(\rho)e^{-a_1\tau} \int_u^{\infty} \frac{e^{-x}}{(\pi x)^{1/2}} erfc\left\{\frac{\eta u^{1/2}}{\left[x(x-u)\right]^{1/2}}\right\} \tag{3.115}$$

$$C_2 = F(\rho)e^{-a_1\tau} \int_u^{\infty} \frac{e^{-x}}{(\pi x)^{1/2}} erfc\left\{\frac{\eta u^{1/2}}{\left[x(x-u)\right]^{1/2}}\right\} erfc(u_1)dx, \tag{3.116}$$

where

$$u = \frac{m^2}{4\pi} \text{ and } u_1 = \frac{\xi}{m}\left(\frac{xu}{x-u}\right)^{1/2}.$$

Solutions for Nondecay Boundary Condition
The Laplace transform of the nondecay boundary is

$$G_1(\rho_0, p) = \frac{1}{p}.$$

G_1 and G_2 in the Laplace domain with the above condition are

$$G_1 = \frac{1}{p}\exp\left(\frac{\rho-\rho_0}{2}\right)\frac{Ai(\beta^{1/3}y)}{Ai(\beta^{1/3}y_0)} \tag{3.117}$$

$$G_2 = \frac{1}{p}\exp\left(\frac{\rho-\rho_0}{2}\right)\exp\left[-\xi(p+a_1)^{1/2}\right]\frac{Ai(\beta^{1/3}y)}{Ai(\beta^{1/3}y_0)}. \tag{3.118}$$

For large p, using (3.115) we get

$$G_1 = \frac{F(\rho)}{p}\exp(-m\beta^{1/2}) \tag{3.119}$$

$$G_2 = \frac{F(\rho)}{p}\exp\left(-m\beta^{1/2}\right)\exp\left[-\xi\left(r+a_1\right)^{1/2}\right]. \tag{3.120}$$

The inversion can be carried out using

$$\mathcal{L}^{-1}\left\{\exp\left[-m\left(p+\alpha p^{1/2}\right)^{1/2}\right]\right\} = \frac{m\alpha^3}{4\pi}\int_0^{\alpha^2\tau}\frac{g(x)}{\left(\alpha^2\tau-x\right)^{3/2}}dx \tag{3.121}$$

where

$$g(x) = \frac{1}{x^{1/2}}\exp\left(-\frac{x^2}{4\left(\alpha^2\tau-x\right)}-\frac{m^2\alpha^2}{4x}\right).$$

Applying the shifting theorem to (3.122), we obtain

$$\mathcal{L}^{-1}\left[\exp\left[-m\beta^{1/2}\right]\right] = \exp\left(-\alpha_1\tau\right)\frac{m\alpha^3}{4\pi}\int_0^{\alpha^2\tau}\frac{g(x)}{\left(a^2\tau-x\right)^{3/2}}dx. \tag{3.122}$$

Using the convolution theorem, we get

$$\mathcal{L}^{-1}\left[\frac{1}{p}\exp\left[-m\beta^{1/2}\right]\right] = \frac{m\alpha^3}{4\pi}\int_0^{\tau}\exp\left(-\alpha_1 k\right)\int_0^{\alpha^2 k}\frac{g(x)}{\left(\alpha^2 k-x\right)^{3/2}}dxdk, \tag{3.123}$$

which can be expressed as

$$\mathcal{L}^{-1}\left[\frac{1}{p}\exp\left[-m\beta^{1/2}\right]\right] = \frac{m\alpha}{\pi}\int_0^{a^2\tau}\frac{g(x)}{x}dx\int_{u'}^{\infty}\exp\left[-k^2-\frac{a^2}{k^2}\right]dk,$$

where

$$u' = \frac{x}{2\left(\alpha^2\tau-x\right)^{1/2}} \text{ and } a = \frac{\alpha_1^{1/2}x}{2\alpha}. \tag{3.124}$$

The inner integral of above is[30]

$$\begin{aligned}\int_{u'}^{\infty}\exp\left(-k^2-\frac{a^2}{k^2}\right)dk &= \frac{\pi^{1/2}}{4}\left[e^{2a}erf\left(k+\frac{a}{k}\right)+e^{-2a}erf\left(k-\frac{a}{k}\right)\right]_{u'}^{\infty} \\ &= \frac{\pi^{1/2}}{4}\left[e^{2a}erfc\left(u'+\frac{a}{u'}\right)+e^{-2a}erfc\left(u'-\frac{a}{u'}\right)\right]\end{aligned} \tag{3.125}$$

where

$$erfc(x)=1-erf(x).$$

Hence

$$\begin{aligned}C_1 &= \frac{F(\rho)}{2(\pi)^{1/2}}\int_u^{\infty}\exp\left(-x-\frac{\delta_1}{x}\right)erfc\left\{\frac{\eta u^{1/2}}{\left[x(x-u)\right]^{1/2}}+\phi\left(\frac{x-u}{x}\right)^{1/2}\right\}\frac{dx}{x^{1/2}} \\ &+\frac{F(\rho)}{2(\pi)^{1/2}}\int_u^{\infty}\exp\left(-x-\frac{\delta_2}{x}\right)erfc\left\{\frac{\eta u^{1/2}}{\left[x(x-u)\right]^{1/2}}-\phi\left(\frac{x-y}{x}\right)^{1/2}\right\}\frac{dx}{x^{1/2}},\end{aligned} \tag{3.126}$$

where the parameters $\delta_{1,}$ $\delta_{2,}$ and φ are functions of α_1 which are all 0 when $\alpha_1 = 0$ (no radioactive decay) and the solution is identical to the small time solution obtained by Chen[31].

Using the shifting rule on the result[32]

$$\mathcal{L}^{-1}\left[\exp\left(-\xi p^{1/2}\right)\right]=\frac{\xi}{2(\pi)^{1/2}}\tau^{-3/2}\exp\left(\frac{-\xi^2}{4\tau}\right), \tag{3.127}$$

---

30 Milton Abramowitz and Irene A. Stegun(1972). Handbook of Mathematical Functions With Formulas, Graphs, and Mathematical Tables, p. 306. https://ia600609.us.archive.org/14/items/HandbookOfMathematicalFunctions/AMS55.pdf

31 Chen, C. S. (1985). Analytical and approximate solutions to radial dispersion from an injection well into a hydrogeologic unit with simultaneous diffusion into adjacent strata, Water Resour. Res., 2•(8), 1069-1076.

32 Erdelyi, A., W. Magnus, F. Oberhettinger, and F. G. Tricomi (1954). Tablesof Integral Transforms, vol. 1, McGraw-Hill, New York.

we obtain

$$\mathcal{L}^{-1}\left\{\exp\left[-\xi\left(p+a_1\right)^{1/2}\right]\right\}=\frac{\xi}{2\left(\pi\right)^{1/2}}\tau^{-3/2}\exp\left(\frac{-\xi^2}{4\tau}-a_1\tau\right). \tag{3.128}$$

Applying the convolution theorem to (3.127) and (3.129) yields the Laplace inverse of (3.121) for small time periods as

$$\begin{aligned} C_2 &= \frac{F(\rho)\xi}{4\pi}\int_0^{\tau}(\tau-k)^{-3/2}\exp[h_1(k)]\int_{\bar{u}}^{\infty}\frac{1}{x^{1/2}}h_2(x,\delta_1)erfc\{h_3(x,\bar{u})+\bar{\phi}(x_1)\}dxdk \\ &+\frac{F(\rho)\xi}{4\pi}\int_0^{\tau}(\tau-k)^{-3/2}\exp[h_1(k)]\int_{\bar{u}}^{\infty}\frac{1}{x^{1/2}}h_2(x,\delta_2)erfc\{h_3(x,\bar{u})-\bar{\phi}(x_1)\}dxdk \end{aligned} \tag{3.129}$$

where

$$h_1(k)=\frac{-\xi^2}{4(\tau-k)}-\alpha_1(\tau-k),\ h_2(x,\delta)=\exp\left(-x-\frac{\delta}{x}\right)$$

$$h_3(x,w)=\frac{\eta w^{1/2}}{\left[x(x-w)\right]^{1/2}}\ \text{and}\ x_1=\left(\frac{x-\bar{u}}{x}\right)^{1/2}.$$

Example 3.7

In this example[33], the radial convection-dispersive equation with slug injection is solved for the case of velocity dependent dispersion coefficient

The flow model describing the overall transport and mixing of incompressible fluids flowing through a porous medium in the presence of miscible displacement of one fluid by another is

$$\frac{\partial C}{\partial t} + v \cdot \nabla C = \nabla \cdot (D \nabla C) \tag{3.130}$$

The boundary conditions at the well bore are

$C(r_w, t) = C_{in}(t)$ or $vC(r_w, t) - D\nabla C(r_w, t) = vC_{in}(t)$.

The equation of motion of the radial dispersion problem is

$$\frac{\partial C}{\partial t} + v_r \frac{\partial C}{\partial r} = \frac{1}{r}\frac{\partial}{\partial r}\left(r \alpha v_r \frac{\partial C}{\partial r} \right) \tag{3.131}$$

where

$D = \alpha v_r$

$\alpha =$ dispersivity.

The initial and boundary conditions are

$$\begin{aligned} &C(r,0) = 0 \\ &C(r,t) \xrightarrow{r\to\infty} 0 \\ &2\pi hr \left(v_r C(r,t) - \alpha v_r \frac{\partial C}{\partial r}(r,t) \right)\bigg|_{r=r_w} = iC_{in} \end{aligned} \tag{3.132}$$

where v is the velocity, i is the volumetric rate of input, r_w is the well bore radius, h is the thickness of formation, and

$$v_r = \frac{i}{2\pi hr}.$$

33 DHE Tang and DW Peaceman (1987). New analytical and numerical solutions for the radial convection-dispersion problem. SPE Reservoir Engineering, Vol. 2, pp.343–359.

Without loss of generality, we will consider only the continuous injection case ($C_{in}(t)= 1$) since the solution for the slug injection case

$$C_{in}(t)=\begin{cases}1 & 0<t\le t_a\\ 0 & \text{otherwise}\end{cases}$$

can be obtained from the former by using superposition.
The dimensionless time and radius are

$$t_D=\frac{it}{2\pi h\alpha^2}$$

$$r_D=\frac{r}{\alpha}.$$

Then we can write equation (3.132) as

$$\frac{\partial^2 C}{\partial r_D^2}-\frac{\partial C}{\partial r_D}=r_D\frac{\partial C}{\partial t_D} \tag{3.133}$$

with initial and boundary conditions

$$\begin{aligned}&C(r_D,0)=0\\ &C(r_{Dw},t)-\frac{\partial C}{\partial_{r_D}}(r_{Dw},t)=C_{in}(t)=1\\ &C(r_D,t)\xrightarrow{r_D\to\infty}0.\end{aligned} \tag{3.134}$$

The Laplace transform of equation (3.134) is

$$s\mathcal{L}\{C\}+\frac{1}{r_D}\frac{d}{dr_D}\mathcal{L}\{C\}=\frac{1}{r_D}\frac{d^2}{dr_D^2}\mathcal{L}\{C\} \tag{3.135}$$

The Laplace transforms of the boundary conditions in (3.135) are

$$\begin{aligned}&\mathcal{L}\{C\}-\frac{d}{dr_D}\mathcal{L}\{C\}=\mathcal{L}\{C_{in}\}=\frac{1}{s}\\ &\mathcal{L}\{C\}(r_D,s)\xrightarrow{r_D\to\infty}0.\end{aligned} \tag{3.136}$$

The solution of the above boundary value problem (3.136) and (3.137) is

$$\mathcal{L}\{C\}=\frac{1}{s}\exp\left(\frac{r_D-r_{Dw}}{2}\right)\left(\frac{Z}{Z_w}\right)^{1/2}\frac{2K_{1/3}(\zeta)}{K_{1/3}(\zeta_w)+2(sZ_w)^{1/2}K_{2/3}\zeta_w} \tag{3.137}$$

where

$$Z=r_D+\frac{1}{4s}$$

$$Z_w=r_{Dw}+\frac{1}{4s}$$

$$\zeta=\frac{2}{3}s^{1/2}Z^{3/2}$$

$$\zeta_w=\frac{2}{3}s^{1/2}Z_w^{3/2},$$

and $K_{1/3}$ and $K_{2/3}$
are modified Bessel functions of the first kind.

Case with $r_w = 0$ and $D\equiv$ constant.

The analytical solution for the constant diffusivity case can be shown to be

$$C(r,t)=\frac{\Gamma\left(\frac{i}{4\pi hD},\frac{r^2}{4D_t}\right)}{\Gamma\left(\frac{i}{4\pi hD}\right)} \tag{3.138}$$

where Γ in the numerator is the incomplete gamma function. Then

$$\frac{\partial C}{\partial r}=-\frac{2}{\Gamma\left(\frac{i}{4\pi hD}\right)}\exp\left(-\frac{r^2}{4Dt}\right)r^{2a-1}(4Dt)^{-a},\text{ with } a=\frac{i}{4\pi hD}. \tag{3.139}$$

We can also show that

$$\lim_{s\to 0}\frac{\partial C}{\partial r}=\begin{cases}-\infty & \text{if } \dfrac{i}{2\pi hD}<1\\ -(\pi Dt)^{-1/2} & \text{if } \dfrac{i}{2\pi hD}=1\\ 0 & \text{if } \dfrac{i}{2\pi hD}>1.\end{cases}\tag{3.140}$$

The above equations imply that the rate caused by dispersion becomes greater than or equal to that by injection. This situation, although feasible mathematically, is unlikely to happen from a physical standpoint, and conclude that

$$\lim_{s\to 0}\frac{\partial C}{\partial r}=0$$

for this case.

Case with $r_w = 0$ and $D\equiv \alpha v_r$

The analytical solution of the system with $r_{Dw}=0$ is

$$C(r_D,t_D)=1-(N_1+N_2)\exp\left(\frac{r_D}{2}\right)\tag{3.141}$$

where N_1 and N_2 involve various Bessel functions in the integrands.

Example 3.8

In this example, the convective-dispersive transport equation with a velocity dependent dispersion coefficient is solved.[34] The solution describes the spatial and temporal distribution of a dissolved substance (tracer) introduced at a well.

The steady state seepage velocity at radius r

is given by the continuity equation

$$v = \frac{Q}{2\pi br\theta} = \frac{A}{r} \tag{3.142}$$

where

$A = \dfrac{Q}{2\pi b\theta}$

$Q =$ constant injection rate

$b =$ distance between the two horizontal permeable beds

$\theta =$ uniform porosity.

The general equation for the distribution of the tracer is

$$\frac{\partial c}{\partial t} + v\frac{\partial c}{\partial r} = vd\frac{\partial^2 c}{\partial r^2} + \frac{D_m}{r}\frac{\partial}{\partial r}\left(r\frac{\partial c}{\partial r}\right) \tag{3.143}$$

where

$c =$ tracer concentration

$v =$ seepage velocity along a streamline

$r =$ radial distance from the center of the well

$r_0 =$ well radius

$D_m =$ molecular diffusion coefficient

$d =$ dispersivity.

Because $D >> D_m$ at small and intermediate distances, (3.144) can be simplified to

---

[34] D.H. Tang and D.K. Babu (1979). Analytical Solution of a Velocity Dependent Dispersion Problem. Water Resources Research, Vol. 15, pp. 1471-1478.

$$\frac{\partial c}{\partial t}+\frac{A}{r}\frac{\partial c}{\partial r}=\frac{Ad}{r}\frac{\partial^2 c}{\partial r^2}. \tag{3.144}$$

Initial and boundary conditions are

$$\begin{aligned} &c(r,0)=0 \text{ for } r\geq r_0 \\ &c(r_0,t)=\ c_0 \text{ for } t>0 \\ &c(r,t)\xrightarrow{r\to\infty}0 \text{ for } t>0. \end{aligned} \tag{3.145}$$

Applying the Laplace transform to (3.145)and (3.146), we get

$$rp\bar{c}+A\frac{d\bar{c}}{dr}=D_0\frac{d^2\bar{c}}{dr^2} \tag{3.146}$$

$$\begin{aligned} &\bar{c}(r_0,p)=c_0/p \\ &\bar{c}(r,p)\xrightarrow{r\to\infty}0 \end{aligned} \tag{3.147}$$

where

$$\begin{aligned} &\bar{c}(r,p)=\int_0^\infty \exp(-pt)c(r,t)dt \\ &D_0=Ad. \end{aligned}$$

Equation (3.147) can now be written as

$$\frac{d^2\bar{c}}{dr^2}-\frac{A}{D_0}\frac{d\bar{c}}{dr}-\frac{pr}{D_0}\bar{c}=0. \tag{3.148}$$

Using the substitution

$$\begin{aligned} &z=r+(\alpha/p) \\ &\alpha=A^2/4D_0 \end{aligned}$$

wecan write (3.149) as

$$\frac{d^2\bar{c}}{dz^2}-\frac{A}{D_0}\frac{d\bar{c}}{dz}-\left(\frac{pz}{D_0}-\frac{A^2}{4D_0^2}\right)\bar{c}=0. \tag{3.149}$$

The general solution is given by

$$\bar{c} = z^{1/2}\exp\left(\frac{Az}{2D_0}\right)\left[c_1 I_{1/3}\left(f(p,z)\right) + c_2 K_{1/3}\left(f(p,z)\right)\right]$$
$$f(p,z) = \frac{2}{3}\left(\frac{p}{D_0}\right)^{1/2} z^{3/2} \tag{3.150}$$

where $I_{1/3}$ and $K_{1/3}$ are modified Bessel function of the first kind and second kind, respectively, andc_1, c_2 are constants. Since $\bar{c}$ is bounded, c_1 = 0 and therefore

$$\bar{c} = c_2 z^{1/2}\exp\left(\frac{Az}{2D_0}\right)K_{1/3}\left(f(p,z)\right) \tag{3.151}$$

Substituting $r = r_0$, $\bar{c} = c_0 / p$, $z = r + a/p$, and $z_0 = r_0 = a/p$ we get

$$\bar{c} = \frac{c_0}{p}\left(\frac{z}{z_0}\right)^{1/2}\exp\left(\frac{A(z-z_0)}{2D_0}\right)\frac{K_{1/3}\left(f(p,z)\right)}{K_{1/3}\left(f(p,z_0)\right)}. \tag{3.152}$$

Which can rewritten as

$$\frac{\bar{c}}{c_0} = \frac{c_0}{p}\left(\frac{z}{z_0}\right)^{1/2}\left(\frac{r}{r_0}\right)^{1/2}\left(\frac{g(p,r)}{g(p,r_0)}\right)^{1/2}\frac{K_{1/3}\left(\beta\left(g(p,r)\right)^{3/2}/p\right)}{K_{1/3}\left(\gamma\left(g(p,r_0)\right)^{3/2}/p\right)}\exp\left(\frac{A(z-z_0)}{2D_0}\right)$$
$$g(p,r) = p + (a/r) \tag{3.153}$$

with

$$\beta = \frac{2}{3}\left(\frac{r^3}{D_0}\right)^{1/2}, \quad \gamma = \frac{2}{3}\left(\frac{r_0^3}{D_0}\right)^{1/2}.$$

Defining ξ and y as

$$\xi = \frac{\beta}{p}\left(p + \frac{\alpha}{r}\right)^{3/2} = \frac{2}{3}Y^{3/2}$$
$$Y = \left(\frac{3}{2}\frac{\beta}{P}\right)^{2/3}\left(p + \frac{\alpha}{r}\right) \tag{3.154}$$

we get

$$K_{1/3}(\xi)=\pi\left(\frac{Y}{3}\right)^{-1/2} Ai(Y) \tag{3.155}$$

where Ai is the Airy function.

Finally

$$\frac{\bar{c}}{c_0}=\frac{1}{p}\exp\left(\frac{A(r-r_0)}{2D_0}\right)\left(\frac{r}{r_0}\right)^{1/2}\left(\frac{\gamma}{\beta}\right)^{1/3}\frac{Ai(Y)}{Ai(Y_0)} \tag{3.156}$$

with

$$Y_0=\left(\frac{3}{2}\frac{\gamma}{p}\right)^{2/3}\left(p+\frac{\alpha}{r_0}\right) \tag{3.157}$$

Inversion of (3.157) yields

$$\frac{c}{c_0}=\left(\frac{r}{r_0}\right)^{1/2}\exp\left(\frac{A(r-r_0)}{2D_0}\right)\frac{1}{2\pi i}\int_{\gamma-i\infty}^{\gamma+i\infty}\frac{e^{pt}}{p}\left(\frac{h(p,r)}{h(p,r_0)}\right)^{1/2}\frac{K_{1/3}\left\{\left(\frac{\beta}{p}\right)(h(p,r))^{3/2}\right\}}{K_{1/3}\left\{\left(\frac{\gamma}{p}\right)(h(p,r_0))^{3/2}\right\}}dp \tag{3.158}$$

where

$h(p,r)=p+\alpha/r.$

The solution of (3.159) can be expressed as

$$\frac{c}{c_0}=1\text{-}\left(\frac{r}{r_0}\right)^{1/2}\exp\left(\frac{A(r-r_0)}{2D_0}\right)(I_1+I_2+I_3) \tag{3.159}$$

where

$$I_1=\frac{1}{2}\int_0^{(\alpha/r)^{1/2}}\frac{2\exp(-\rho^2 t)}{\rho}\left(\frac{(a/r)-\rho^2}{(a/r_0)-\rho^2}\right)^{1/2}d\rho$$
$$\times\left\{\frac{K_{1/3}(\arg\beta)I_{1/3}(\arg\gamma)-K_{1/3}(\arg\gamma)I_{1/3}(\arg\beta)}{\frac{1}{4}(K_{1/3}(\arg\gamma))^2+\left[(3^{1/2}/2)K_{1/3}(\arg\gamma)+\pi I_{1/3}(\arg\gamma)\right]^2}\right\} \tag{3.160}$$

with

$$\arg\beta=\beta\left((a/r)-\rho^{2}\right)^{3/2}/\rho^{2},\ \arg\gamma=\gamma\left((a/r_{o})-\rho^{2}\right)^{3/2}/\rho^{2}$$

$$I_{2}=-\frac{1}{2}\int_{(\alpha/r)^{1/2}}^{(\alpha/r_{0})^{1/2}}\frac{2\exp\left(-\rho^{2}t\right)}{\rho}\left(\frac{\rho^{2}-(\alpha/r)}{(\alpha/r_{0})-\rho^{2}}\right)^{1/2}d\rho$$

$$\times\frac{[-K_{1/3}(\arg\gamma)J_{1/3}(\arg\beta)+\pi I_{1/3}(\arg\gamma)(-(3^{1/2}/2)J_{1/3})(\arg\beta)+\frac{1}{2}Y_{1/3}(\arg\beta))]}{\left[\frac{1}{4}\left(K_{1/3}(\arg\gamma)\right)^{2}\right]+\left((3^{1/2}/2)K_{1/3}(\arg\gamma)+\pi I_{1/3}(\arg\gamma)\right)^{2}]},\quad(3.161)$$

$$\arg\beta=\frac{\beta}{\rho^{2}}\left(\rho^{2}-\frac{a}{r}\right)^{3/2},\ \arg\gamma=\frac{\gamma}{\rho^{2}}\left(\frac{a}{r_{o}}-\rho^{2}\right)^{3/2}$$

$$I_{3}=\frac{1}{\pi}\int_{(a/r_{o})^{1/2}}^{\infty}\frac{2\exp\left(-\rho^{2}t\right)}{\rho}\left(\frac{\rho^{2}-(a/r)}{\rho^{2}-(a/r_{o})}\right)^{1/2}d\rho$$

$$\times\frac{\left\{J_{1/3}(\arg\gamma)Y_{1/3}(\arg\beta)-J_{1/3}(\arg\beta)Y_{1/3}(\arg\gamma)\right\}}{\left[\left(J_{1/3}(\arg\gamma)\right)^{2}+\left(Y_{1/3}(\arg\gamma)\right)^{2}\right]},\quad(3.162)$$

$$\arg\beta=\frac{\beta}{\rho^{2}}\left(\rho^{2}-\frac{\alpha}{r}\right)^{3/2},\ \arg\gamma=\frac{\gamma}{\rho^{2}}\left(\rho^{2}-\frac{\alpha}{r_{o}}\right)^{3/2}$$

The expression (3.160) for small values of $\alpha t/r_{o}$ and large values of $\beta t^{-1/2}$ is very difficult to evaluate. The asymptotic expressions for small and large times are :

$$\frac{c}{c_{o}}\approx\left(\frac{r}{r_{o}}\right)^{-1/4}\exp\left(\frac{A(r-r_{o})}{2D_{o}}\right)\operatorname{erfc}\left(\frac{\beta-\gamma}{2t^{1/2}}\right)\text{ if }(at/r_{o})<<1,\quad(3.163)$$

$$\frac{c}{c_{o}}\approx1-\exp\left(\frac{A(r-r_{o})}{2D_{o}}\right)\times\left(I_{4}+I_{5}\right)/\pi\text{ for }\beta t^{-1/2}>>10,\quad(3.164)$$

where

$$I_4 = \int_{\alpha/r}^{\alpha/r_0} \left(\frac{4\alpha + (r - r_0)x}{4\alpha x} \right) \exp\left(-tx - \frac{\gamma}{x}\left(\frac{\alpha}{r_0} - x \right)^{3/2} \right) \sin\left(\frac{\beta}{x}\left(x - \frac{\alpha}{r} \right)^{3/2} \right) dx \tag{3.165}$$

$$I_5 = \int_{\alpha/r_0}^{\infty} \left(\frac{4\alpha + (r - r_0)x}{4\alpha x} \right) \exp(-tx) \sin\left(\frac{(\beta(x - (\alpha/r))^{3/2} - \gamma(x - (\alpha/r_0))^{3/2}}{x} \right) dx \tag{3.166}$$

Example 3.9

This example (Chen[35], 1987) is concerned with the radial dispersion problem under the Cauchy boundary condition at injection well. The transport mechanism considered are: radial advection and longitudinal dispersion; the effect of molecular diffusion is neglected.

Let C = dimensionless concentration =concentration/ C_o

τ = dimensionless = At/α^2

ρ = dimensionless radial distance = r/α

ρ_0 = dimensionless well radius = r_w/α

α = longitudinal dispersivity for the aquifer, assumed constant.

The mathematical model in terms of dimensionless variables is for this case is

$$\frac{\partial^2 C}{\partial \rho^2} - \frac{\partial C}{\partial \rho} = \rho \frac{\partial C}{\partial \rho} \tag{3.167}$$

$$\begin{aligned} &C(\rho,0) = 0 \\ &C - \frac{\partial C}{\partial \rho} = 1 \text{ at } \rho_0 \\ &C(\rho,\tau) = 0 \text{ as } \rho \to \infty \end{aligned} \tag{3.168}$$

Taking the Laplace transform of (3.168) and (3.169), we obtain

$$\frac{d^2 G}{d\rho^2} - \frac{dG}{d\rho} - \rho p G = 0 \tag{3.169}$$

$$\begin{aligned} &G - \frac{dG}{d\rho} = \frac{1}{p} \text{ at } \rho_0 \\ &G(\rho,p) = 0 \text{ as } \rho \to \infty \end{aligned} \tag{3.170}$$

---

35 C. S. Chen (1987). Analytical Solutions for Radial Dispersion With Cauchy Boundary at Injection Well. Water Resources Research. Vol. 23, pp 1217-1224.

where G is the Laplace transform of C given by

$$G(p)=\int_0^\infty e^{-p\tau}C(\rho,\tau)\,d\tau.$$

The solution to equations (3.170) and (3.171) is

$$G=\exp\left(\frac{\rho-\rho_0}{2}\right)\frac{1}{p}\frac{Ai(Y)}{(1/2)Ai(Y_0)-p^{1/3}Ai'(Y_0)} \tag{3.171}$$

where

$$\begin{aligned} Y&=p^{1/3}\left(\rho+\frac{1}{4p}\right)=\frac{1+4\rho p}{4p^{2/3}}\\ Y_0&=p^{1/3}\left(\rho_0+\frac{1}{4p}\right)=\frac{1+4\rho_0 p}{4p^{2/3}} \end{aligned} \tag{3.172}$$

The Airy function $Ai(Y)$ satisfies the following relationship

$$Ai''(Y)=YAi(Y) \tag{3.173}$$

where $Ai'(Y)$ and $Ai''(Y)$ are the first and second order derivatives of $Ai(Y)$, respectively.

Asymptotic solution for small τ.

The small τ solution for C corresponds to large p solution for G. For large $|p|$, G can be expressed as

$$G=\exp\left(\frac{\rho-\rho_0}{2}\right)(\rho_0\rho)^{-1/4}\frac{\exp\left(-m\sqrt{p}\right)}{\left(\sqrt{\rho_0}/2+\sqrt{p}\right)}\left(\frac{1}{p}\right) \tag{3.174}$$

where $m=2\left(\rho^{3/2}-P_0^{3/2}\right)/3$.

In the above, we have used the following approximations (Abramowitz and Stegun[36]):

$$
\begin{aligned}
Ai(z) &\simeq \frac{1}{2\sqrt{\pi}} z^{-1/4} \exp\left(-\frac{2}{3} z^{3/2}\right) \\
Ai'(z) &\simeq -\frac{1}{2\sqrt{\pi}} z^{1/4} \exp\left(-\frac{2}{3} z^{3/2}\right)
\end{aligned}
\tag{3.175}
$$

The Laplace inverse of (3.175) is:

$$
\begin{aligned}
C(\rho_1, \tau \ll 1) = 2\exp\left(\frac{\rho-\rho_0}{2}\right)\left(\frac{\rho_0}{\rho}\right)^{1/4} \times \\
\left\{ \operatorname{erfc}\left(\frac{m}{2\sqrt{\tau}}\right) - \exp\left[\frac{m}{2\sqrt{\rho_0}} + \frac{\tau}{4\rho_0}\right] \times \operatorname{erfc}\left[\frac{m}{2\sqrt{\tau}} + \frac{1}{2}\sqrt{\frac{\tau}{\rho_0}}\right] \right\}
\end{aligned}
\tag{3.176}
$$

The concentration variation at ρ_0 for small τ is

$$
C_0(\rho_0, \tau) = 2\left(1 - \exp\left(\frac{\tau}{4\rho_0}\right) \operatorname{erfc}\left(\frac{1}{2}\sqrt{\frac{\tau}{\rho_0}}\right)\right)
\tag{3.177}
$$

Exact solution

The Laplace inverse of G in (3.175) is

$$
C = \frac{1}{2\pi i} \int_{\gamma - i\infty}^{\gamma + i\infty} \exp\left(\frac{\rho-\rho_0}{2}\right) \frac{e^{pt}}{p} \frac{A_i(Y)}{A_i(Y_0)/2 - p^{1/3} A_i'(Y_0)} dp,
\tag{3.178}
$$

where $\gamma > 0$ is chosen so that all of the singularities of the integrand lie to the left of the vertical line $\gamma - i\infty$ to $\gamma + i\infty$. The Cauchy integral theory yields

$$
C = 1 - \frac{4}{\pi} \exp\left(\frac{\rho-\rho_0}{2}\right) \int_0^{\infty} \frac{e^{-x^2\tau}}{x} \frac{Ai(\eta) f_1 - Bi(\eta) f_2}{f_1^2 + f_2^2} dx
\tag{3.179}
$$

---

[36] Abramowitz, M. and Stegun, I.A. (1970). Handbook of Mathematical Functions, Applied Math Series, 55, National Bureau of Standards, p. 475.

where

$$
\begin{aligned}
f_1(x) &= Bi(\eta_0) + 2x^{2/3} Bi'(\eta_0) \\
f_2(x) &= Ai(\eta_0) + 2x^{2/3} Ai'(\eta_0) \\
\eta(x) &= \frac{1-4\rho x^2}{4x^{4/3}} \\
\eta_0(x) &= \frac{1-4\rho_0 x^2}{4x^{4/3}}
\end{aligned}
\tag{3.180}
$$

where $Bi(z)$ denotes unbounded Airy function and $Bi'(z)$ is its derivative of first order. For large $|z|$ we can use the following approximations[37]

$$
\begin{aligned}
Bi(z) &\simeq \frac{1}{\sqrt{\pi}} z^{-1/4} \exp\left(\frac{2}{3} z^{3/2}\right) \\
Bi'(z) &\simeq \frac{1}{\sqrt{\pi}} z^{1/4} \exp\left(\frac{2}{3} z^{3/2}\right)
\end{aligned}
\tag{3.181}
$$

The Wronskian of $Ai(z)$ and $Bi(z)$ is

$$
W(Ai(z), Bi(z)) = Ai(z)Bi'(z) - Ai'(z)Bi(z) = \frac{1}{\pi} \tag{3.182}
$$

showing that these two functions are linearly independent.

37 Abramowitz and Stegun, 1970, p. 449)

Example 3.10

For remediation of a superfund site with contaminated ground water the method of "pump and treat" is commonly used by the U.S. EPA[38]. In this example (Chen and Woodside[39], 1988), a mathematical model for aquifer decontamination by pumping is developed. The pumping well with a constant flow rate is the sink located at the center of the plume. The transport mechanism is assumed to be advection and longitudinal mechanical dispersion.

The groundwater flow caused by pumping is assumed to be radially symmetric and convergent. It can be described by

$$v(r) = -\frac{Q}{2\pi bnr} \qquad r_w \le r < \infty \tag{3.183}$$

where

$v =$ the groundwater velocity
$b =$ the constant aquifer thickness
$Q =$ the constant pumping rate
$n =$ the constant aquifer porosity
$r =$ the radial distance measured from the well radius r_w

The negative sign in (3.184) indicates that the radial flow is convergent to the withdrawal well. The governing equation for the radial dispersion on the groundwater flow field can be formulated in a dimensionless from as

$$\frac{\partial^2 C}{\partial \rho^2} + \frac{\partial C}{\partial \rho} = \rho \frac{\partial C}{\partial \tau} \qquad \rho_0 \le \rho < \infty \tag{3.184}$$

---

38 Groundwater Pump and Treat Systems: Summary of Selected Cost and Performance Information at Superfund-financed Sites. EPA 542-R-01-021b, December 2001(http://epa.gov/tio/download/remed/542r01021b.pdf)

39 C.S. Chen and G. D. Woodside(1988). Analytical Solution for Aquifer Decontamination by Pumping. Water Resources Research. Vol. 24, pp 1329-1338.

where

$C =$ the concentration

$\rho =$ the dimemsionless radial distance r/d

$\rho_0 =$ the dimensionless well radius defined by r_w/d

$\tau =$ the dimensionless time defined by At/d^2

$A = Q/2\pi bn$

$d =$ the constant longitudinal dispersivity

The initial conditions are

$$\begin{aligned} C(\rho,0) &= f(\rho) = C_0 - m_1(\rho-\rho_0),\ \rho_0 \le \rho \le \rho \\ C(\rho,0) &= f(\rho) = C_0 - m_1(\rho-\rho_0) - m_2(\rho-\rho_1),\ \rho_1 \le \rho \le \rho_2 \\ C(\rho,0) &= f(\rho) = C_0 - m_1(\rho-\rho_0) - \sum_{i=1}^{n-1} m_{i+1}(\rho_{i+1}-\rho_i) = 0,\ \rho_n \le \rho < \infty \end{aligned} \tag{3.185}$$

where C_0 is the peak concentration, and $m_1, m_2, \ldots, m_n$ are the absolute values of slopes of the straight lines:

$$m_i = \left|(C_i - C_{i-1})/(\rho_i - \rho_{i-1})\right|,\ i = 1,2,\ldots,n$$

and C_i is the known concentration at a distance ρ_i.

The boundary condition are

$$\begin{aligned} &\frac{\partial C}{\partial \rho} = 0 \ \text{ at } \rho = \rho_0 \\ &C(\rho,\tau) \to 0 \text{ as } \rho \to \infty \end{aligned} \tag{3.186}$$

Applying the Laplace transform with respect to τ, we obtain

$$\frac{d^2G}{d\rho^2} + \frac{dG}{d\rho} - \rho p G = -\rho f(\rho) \tag{3.187}$$

$$\begin{aligned} &\frac{dG}{d\rho} = 0 \ \text{ at } \rho = \rho_0 \\ &G(\rho,p) \to 0 \ \text{ as } \ \rho \to \infty \end{aligned} \tag{3.188}$$

where

$$G(\rho,p)=\int_0^\infty \bar{e}^{p\tau}C(\rho,\tau)d\tau$$

The self-adjoint form of (3.188) and (3.189) is

$$\frac{d^2G_1}{d\rho^2}-\left(\rho p+\frac{1}{4}\right)G_1=-F(\rho) \tag{3.189}$$

where

$$G_1=e^{p/2}G$$
$$F(\rho)=\rho e^{p/2}f(\rho)$$

For $F(\rho)$ to be finite, we require $f(\rho)\sim o\left(\rho e^{\rho/2}\right)$.

The boundary conditions transform to

$$\begin{aligned}&\frac{dG_1}{d\rho}-\frac{1}{2}G_1=0 \text{ at } \rho=\rho_0\\&G_1\sim o\left(e^{-\rho/2}\right) \text{ as } \rho\to\infty\end{aligned} \tag{3.190}$$

The equations (3.190) and (3.191) forma regular Sturm-Louiville problem, whose general solution is [Wylie[40], 1975]

$$G_1(\rho,p)=\int_{\rho_0}^\infty g(\rho,p,\xi)F(\xi)d\xi \tag{3.191}$$

where $g(\rho,p,\xi)$ is the Green's function for the problem, with the following properties:

1. $g(\rho,p,\xi)$ satisfies the homogeneous condition of (3.190)
   for $\rho_0\le\rho<\xi$ and $\xi<\rho<\infty$,
2. $g(\rho,p,\xi)$ satisfies the boundary conditions (3.191),
3. $g_{\rho=\xi^+}\Big|=g\Big|_{p=\xi^-}$,
4. $\left.\frac{dg}{d\rho}\right|_{p=\xi^+}=\left.\frac{dg}{d\rho}\right|_{p=\xi^-}=-1$

[40] Wiley, C.R. (1975). Advanced Engineering Mathematics, 4th Ed., McGraw Hill, New York.

The solution to the homogeneous problem for (3.190) in terms of the Airy functions is given by:

$$\begin{aligned} g_1(\rho,p,\xi) &= C_1 Ai(y) + C_2 Bi(y) \quad \rho_0 \le \rho < \xi \\ g_2(\rho,p,\xi) &= C_3 Ai(y) + C_4 Bi(y) \quad \xi \le \rho < \infty \end{aligned} \tag{3.192}$$

where

$$y = p^{1/3}\left(\rho + \frac{1}{4p}\right)$$

and $C_j, j = 1,\ldots,4$ are the coefficients to be determined from the boundary conditions (3.191).

The complete solution is

$$\begin{aligned} g_1(\rho,p,\xi) &= \left(\frac{\pi}{p^{1/3}}\right) Ai(s)\left[Bi(y) - XAi(y)\right] \quad \rho_0 \le \rho < \xi \\ g_2(\rho,p,\xi) &= \left(\frac{\pi}{p^{1/3}}\right) Ai(y)\left[Bi(s) - XAi(s)\right] \quad \xi \le \rho < \infty \end{aligned} \tag{3.193}$$

where

$$X = \frac{p^{1/3}Bi'(y_0) - Bi(y_0)/2}{p^{1/3}Ai'(y_0) - Ai(y_0)/2}, \; s = p^{1/3}\left(\xi + \frac{1}{4p}\right), \text{ and } y_0 = p^{1/3}\left(\rho_0 + \frac{1}{4p}\right).$$

The Laplace domain solution for the problem (3.190) and (3.191) is

$$G(\rho,p) = e^{-\rho/2}\left\{\int_{\rho_0}^{\rho} g_2(\rho,p,\xi)F(\xi)d\xi + \int_{\rho}^{\infty} g_1(\rho,p,\xi)F(\xi)d\xi\right\} \tag{3.194}$$

which can be expressed as

$$G(\rho,p) = \pi e^{-p/2} \times \left\{\int_{\rho_0}^{\rho} H_1F(\xi)d\xi + \int_{\rho_0}^{\infty} H_1F(\xi)d\xi - \int_{\rho_0}^{\infty} H_3F(\xi)d\xi\right\} \tag{3.195}$$

with

$$H_1 = Ai(y)Bi(s)/p^{1/3}$$
$$H_2 = Ai(s)Bi(y)/p^{1/3}$$
$$H_3 = \left(\frac{Ai(y)Ai(s)}{p^{1/3}}\right)\left(\frac{p^{1/3}Bi'(y_0)-\frac{1}{2}Bi(y_0)}{p^{1/3}Ai'(y_0)-\frac{1}{2}Ai(y_0)}\right).$$

The Laplace inversion of (3.196)yields

$$C(\rho,\tau)=\frac{e^{-\rho/2}}{2}\int_{\rho_0}^{\infty}F(\xi)\int_0^{\infty}x^{1/3}e^{-x^2\tau}\left(L-\frac{4JKN+M\left(3J^2-K^2\right)}{J^2+K^2}\right)dxd\xi \tag{3.196}$$

where

$$J=x^{2/3}Ai'(\phi_0)+Ai(\phi_0)/2$$
$$K=x^{2/3}Bi'(\phi_0)+Bi(\phi_0)/2$$
$$L=3Ai(\phi)Ai(\psi)+Bi(\phi)Bi(\psi)$$
$$M=Ai(\phi)Bi(\psi)-Bi(\phi)Bi(\psi)$$
$$N=Ai(\phi)Bi(\psi)+Ai(\psi)Bi(\phi)$$
$$\phi_0(x)=\frac{1-r\rho_0x^2}{4x^{4/3}}$$
$$\phi(x)=\frac{1-4\rho x^2}{4x^{4/3}}$$
$$\psi(x)=\frac{1-4\xi x^2}{4x^{4/3}}.$$

Example 3.11

In this example[41] transport of a contaminant from an injection well into an aquifer with simultaneous diffusion into an adjacent aquitards is investigated. The governing equations for concentration in the aquifer and aquitards with prescribed initial and boundary conditions in dimensionless form are

$$\frac{\partial^2 C_2}{\partial \xi^2} = \frac{1}{\omega_2^2}\frac{\partial C_2}{\partial \tau} \qquad 0 \le \xi < \infty \tag{3.197}$$

$$\begin{aligned} C_2(0,\tau) &= C_1(\rho,\tau) \\ C_2(\infty,\tau) &= 0 \end{aligned} \tag{3.198}$$

and

$$\frac{1}{\rho}\frac{\partial^2 C_1}{\partial \rho^2} - \frac{1}{\rho}\frac{\partial C_1}{\partial \rho} + \frac{\omega_1^2}{\ell}\frac{\partial C_2}{\partial \xi}\bigg|_{\xi=0} = \frac{\partial C_1}{\partial \tau} \qquad \rho_0 \le \rho < \infty \tag{3.199}$$

$$\begin{aligned} C_1(\rho_0,\tau) &= 1 \\ C_1(\infty,\tau) &= 0 \\ C_1(\rho,0) &= C_2(\xi,0) = 0 \end{aligned} \tag{3.200}$$

where

$C_1 =$ normalized contaminant concentration in the main aquifer

$C_2 =$ normalized contaminant concentration in the low permeability aquitard

$\tau =$ dimensionless time

$\xi =$ vertical distance

$\rho =$ radial distance

$\rho_0 =$ well radius

$\ell =$ half aquifer thickness

$\omega_i =$ ratio of diffusion losses to the injection effect $i = 1, 2$.

41 Chen, C.S. (1985). Analytical and Approximate Solutions to Radial Dispersion from an Injection Well to a Geological Unit with Simultaneous Diffusion into Adjacent Strata. Water Resources Research, Vol. 21, No. 8, pp. 1069-1076.

Applying the Laplace transformation to (3.198) – (3.201), we obtain

$$\frac{d^2G_2}{d\xi^2} = \frac{p}{\omega_2^2}G_2 \tag{3.201}$$

$$\begin{aligned} G_2(0,p) &= G_1(\rho,p) \\ G_2(\infty,p) &= 0 \end{aligned} \tag{3.202}$$

and

$$\frac{1}{\rho}\frac{d^2G_1}{d\rho^2} - \frac{1}{\rho}\frac{dG_1}{d\rho} + \frac{\omega_1^2}{\ell}\frac{dG_2}{d\xi}\bigg|_{\xi=0} = pG_1 \tag{3.203}$$

$$\begin{aligned} G_1(\rho_0,p) &= \frac{1}{p} \\ G_1(\infty,p) &= 0 \end{aligned} \tag{3.204}$$

where

$$\begin{aligned} G_1 &= \int_0^\infty e^{-p\tau}C_1(\rho,\tau)d\tau \\ G_2 &= \int_0^\infty e^{-p\tau}C_2(\xi,\tau)d\tau. \end{aligned}$$

The Laplace domain solution of (3.202) and (3.203) is

$$G_2 = \exp\left(-\xi p^{1/2}/\omega_2\right)G_1 \tag{3.205}$$

and the gradient of G_2 at $\xi = 0$ is

$$\frac{dG_2}{d\xi}\bigg|_{\xi=0} = -\frac{p^{1/2}}{\omega_2}G_1. \tag{3.206}$$

Substituting (3.207) in (3.204) we obtain

$$\frac{d^2G_1}{d\rho^2} - \frac{dG_1}{d\rho} - B\rho G_1 = 0 \tag{3.207}$$

where $\beta = p + ap^{1/2}$ and $\alpha = (n_2\omega_2)/n_i\ell$.

In case there is no loss due to diffusion, ω_2 and α becomes zero, *i.e.*, $\beta = p$.

The functions G_1 and G_2 can be expressed in terms of the Airy functionsas

$$G_1 = \frac{1}{p}\exp\left(\frac{y-y_0}{2}\right)\frac{Ai\left(\beta^{1/3}y\right)}{Ai\left(B^{1/3}y_0\right)} \tag{3.208}$$

$$G_2 = \frac{1}{p}\exp\left(\frac{y-y_0}{2}\right)\frac{Ai\left(\beta^{1/3}y\right)}{Ai\left(\beta^{1/3}y_0\right)}\exp\left(-\alpha_1 p^{1/2}\right) \tag{3.209}$$

where $\alpha_1 = \xi / \omega_2$.

Using the approximation to the Airy function

$$Ai(x) \simeq \frac{1}{2(\pi)^{1/2}}(x)^{-1/4}\exp\left(-\frac{2}{3}x^{3/2}\right)$$

For large p, we can express (3.209)-(3.210)as

$$G_1 = \left(\frac{\rho_0}{\rho}\right)^{1/4}\exp\left(\frac{\rho-\rho_0}{2}\right)\left(\frac{\exp\left(-m(p+\alpha p^{1/2})^{1/2}\right)}{p}\right) \tag{3.210}$$

and

$$G_2 = \exp(-\alpha_1 p^{1/2})G_1 \tag{3.211}$$

where $m = \frac{2}{3}\left(\rho^{3/2} - \rho_0^{3/2}\right)$.

The exact solution of (3.211) and (3.212) obtained from inversion is

$$C_1 = F(\rho)\int_u^{\infty}\frac{e^{-x}}{(\pi x)^{1/2}}\,erfc\left(\frac{\eta u^{1/2}}{\left(x(x-u)\right)^{1/2}}\right)dx \tag{3.212}$$

$$C_2 = F(\rho)\int_u^{\infty}\frac{e^{-x}}{(\pi x)^{1/2}}\,erfc\left(\eta\frac{u^{1/2}}{\left(x(x-u)\right)^{1/2}}\right)erfc(u_1)\,dx \tag{3.213}$$

with

$$F(\rho)=(\rho_0/\rho)^{1/4}\exp\left(\frac{\rho-\rho_0}{2}\right),\ \eta=m\alpha/4,\ \text{and } u=m^2/(4\tau)$$

$$u_1=\frac{\alpha_1}{m}\left(\frac{xu}{x-u}\right)^{1/2} \qquad u\le x<\infty.$$

Example 3.12

In this example [42], the method of Laplace transform is applied to find the general solution of the problem of contaminant transport in layered porous media, which is modeled as a quasi-three dimensional advective diffusion equation.

The governing transport equation and boundary conditions in the main aquifer are

$$\frac{\partial c_m}{\partial t}+\frac{V_{xm}}{R_m}\frac{\partial c_m}{\partial x}+\frac{Vz_m}{R_m}\frac{\partial c_m}{\partial z}-\frac{D_{xxm}}{R_m}\frac{\partial^2 c_m}{\partial x^2}-\frac{D_{yym}}{R_m}\frac{\partial^2 c_m}{\partial y^2}-2\frac{D_{xzm}}{R_m}\frac{\partial^2 c_m}{\partial x \partial z} -\frac{D_{zzm}}{R_m}\frac{\partial^2 c_m}{\partial z^2}+\lambda c_m=0 \tag{3.214}$$

where

m = index for the main aquifer

c_m = local concentration

V_{xm} = longitudinal pore velocity

V_{zm} = vertical pore velocity

D_{xxm}, D_{xxm}, D_{xxm}, D_{xxm} = components of the hydrodynamic dispersion coefficients

λ = decay constant

R_m = retardation coefficient

The interface boundary conditions satisfying continuity of concentration in the aquitard boundaries are

$$\begin{gathered}
c_o(x,y,z_b,t)=c_m(x,y,z_b,t)\\
c_u(x,y,z_a,t)=c_m(x,y,z_a,t)\\
\left.\left(n_m c_m V_{zm}-n_m D_{zzm}\frac{\partial c_m}{\partial z}-2D_{xzm}\frac{\partial^2 c_m}{\partial x\partial z}\right)\right|_{z=z_b}=\left.\left(n_o c_o V_{zo}-n_o D_{zzo}\frac{\partial c_o}{\partial z}\right)\right|_{z=z_b}\\
\left.\left(n_m c_m V_{zm}-n_m D_{zzm}\frac{\partial c_m}{\partial z}-2D_{xzm}\frac{\partial^2 c_m}{\partial x\partial z}\right)\right|_{z=z_a}=\left.\left(n_u c_u V_{zu}-n_u D_{zzu}\frac{\partial c_u}{\partial z}\right)\right|_{z=z_a}
\end{gathered} \tag{3.215}$$

42 Yi Tang and Mustafa M. Aral (1992). Contaminant Transport in Layered PorousMedia, 1. General Solution. Water Resources Research, Vol. 28, pp. 1389-1397.

Here the subscripts *o* and *u* refer to the overlying and the underlying aquitards respectively.

Integration of (3.215) with respect to *z* in between aquifer boundaries z_a and z_b and using (3.216) yields

$$\frac{\partial C}{\partial t}+\frac{V_{xm}}{R_m}\frac{\partial C}{\partial x}-\frac{D_{xxm}}{R_m}\frac{\partial^2 C}{\partial x^2}-\frac{D_{yym}}{R_m}\frac{\partial^2 C}{\partial y^2}+\lambda C+\frac{n_u D_{zzu}\left.\frac{\partial c_u}{\partial z}\right|_{z=z_a}-n_o D_{zzo}\left.\frac{\partial c_o}{\partial z}\right|_{z=z_b}}{n_m R_m\left(z_b-z_a\right)}+\frac{n_u c_u V_{zu}\left.\right|_{z=z_a}-n_o c_o V_{zo}\left.\right|_{z=z_b}}{n_m R_m\left(z_b-z_a\right)}=0 \tag{3.216}$$

where c_u and c_o are the concentrations, D_{zzu} and D_{zzo} are the dispersion coefficients, n_u and n_o are the porosities of the underlying and overlying aquitards, and C is the vertically averaged concentration in the main aquifer.

Using the mass balance equation and the continuity of concentration at z_a and z_b we obtain

$$\begin{aligned} n_u V_{zu} &= n_o V_{zo} \\ c_u\left.\right|_{z=z_a} &= c_u\left.\right|_{z=z_b}=C \end{aligned} \tag{3.217}$$

and (3.217) becomes

$$\frac{\partial C}{\partial t}+\frac{V_{xm}}{R_m}\frac{\partial C}{\partial x}-\frac{D_{xxm}}{R_m}\frac{\partial^2 C}{\partial x^2}-\frac{D_{yym}}{R_m}\frac{\partial^2 C}{\partial y^2}+\lambda C+\frac{n_u D_{zzu}\left.\frac{\partial c_u}{\partial z}\right|_{z=z_a}-n_o D_{zzo}\left.\frac{\partial c_o}{\partial z}\right|_{z=z_b}}{n_m R_m\left(z_b-z_a\right)}=0 \tag{3.218}$$

The boundary and initial conditions of (3.219) are given as

$$\begin{aligned} &C(0,y,t)=C_1(y)\exp(r_1 t) \\ &C(L_{1,}y,t)=C_2(y)\exp(r_2 t) \\ &\frac{\partial C}{\partial y}(x,0,t)=0 \\ &\frac{\partial C}{\partial y}(x,L_2,t)=0 \\ &L(x,y,0)=0 \end{aligned} \tag{3.219}$$

where r_1, r_2 are constants and $r_1 = r_2 = 0$ correspond to a step function boundary condition.

The solution of (3.219) involves the coupling equations

$$\frac{\partial c_k}{\partial t} + \frac{V_{zk}}{R_k}\frac{\partial c_k}{\partial z} + \frac{D_{zzk}}{R_k}\frac{\partial^2 c_k}{\partial z^2} + \lambda c_k = 0,\ k = u,\ o \tag{3.220}$$

The boundary and initial conditions for the aquitards are

$$\begin{aligned} c_u(x,y,z_a,t) &= c_o(x,y,z_b,t) = C \\ c_u(x,y,-\infty,t) &= c_o(x,y,\infty,t) = 0 \\ c_u(x,y,z,0) &= c_o(x,y,z,0) = 0 \end{aligned} \tag{3.221}$$

Applying the Laplace transformation with respect to t, equations (3.219) and (3.220)become

$$s\bar{C} + V_{xm}\frac{\partial \bar{C}}{\partial x} - \frac{D_{xxm}}{R_m}\frac{\partial^2 \bar{C}}{\partial x^2} - \frac{D_{yym}}{R_m}\frac{\partial^2 \bar{C}}{\partial y^2} + \lambda\bar{C} + \frac{n_u D_{zzu}\left.\frac{\partial \bar{c}_u}{\partial z}\right|_{z=z_a} - n_o D_{zzo}\left.\frac{\partial \bar{c}_o}{\partial z}\right|_{z=z_b}}{n_m R_m (z_b - z_a)} = 0 \tag{3.222}$$

$$\begin{aligned} \bar{C}(0,y,s) &= \frac{C_1}{s-r_1} \\ \bar{C}(L_1,y,s) &= \frac{C_2}{s-r_2} \\ \frac{\partial \bar{C}(x,0,s)}{\partial y} &= 0 \\ \frac{\partial \bar{C}(x,L_2,s)}{\partial y} &= 0 \end{aligned} \tag{3.223}$$

and equations (3.221) and (3.222) for the aquitards transform to

$$s\bar{c}_k + \frac{V_{zk}}{R_k}\frac{\partial \bar{c}_k}{\partial z} - \frac{D_{zzk}}{R_k}\frac{\partial^2 \bar{c}_k}{\partial z^2} + \lambda\bar{c}_k = 0,\ k = u,\ o \tag{3.224}$$

$$\begin{aligned} \bar{c}_u(x,y,z_a,s) &= \bar{c}_o(x,y,z_b,s) = \bar{C}(s) \\ \bar{c}_u(x,y,-\infty,s) &= \bar{c}_o(x,y,\infty,s) = 0 \end{aligned} \tag{3.225}$$

where

$$\bar{c} = \mathcal{L}\left[c(x,y,t)\right] = \int_0^\infty \exp(-st)c(x,y,t)dt$$

and $\mathcal{L}$ is the Laplace transform symbol.

The general solution of (3.221) depends on the solutions of (3.223) and (3.224). The solutions of the latter equations are given by

$$\begin{aligned}\bar{c}_k(x,y,z,s) = A_K(s)\exp&\left(\frac{V_{zk} + \left(V_{zk}^2 + 4D_{zzk}R_k\lambda + 4D_{zzk}R_k s\right)^{1/2}}{2D_{zzk}} z\right)\\ +B_k(s)\exp&\left(\frac{V_{zk} + \left(V_{zk}^2 - 4D_{zzk}R_k\lambda + 4D_{zzk}R_k s\right)^{1/2}}{2D_{zzk}} z\right), k = u,\ o\end{aligned} \tag{3.226}$$

The constants A_k and B_k are determined from the boundary conditions for the underlying aquitard as follows:

$$\begin{aligned}&z \to \infty,\ \bar{c}_u(x,y,z,s) = 0 \Rightarrow B_u(s) = 0\\ &z = z_a,\ \bar{c}_u(x,y,z,s) = \bar{C}(x,y,s)\\ &\Rightarrow A_u(s) = \bar{c}\exp\left(-\frac{V_{zu} - \left(V_{zu}^2 + 4D_{zzu}R_u\lambda + 4D_{zzu}R_u s\right)^{1/2}}{2D_{zzu}} z_a\right)\end{aligned} \tag{3.227}$$

Substituting the values of the constants in (3.227) we obtain, for the underlying aquitard

$$\bar{c}_u(x,y,z,s) = \bar{C}\exp\left(\frac{V_{zu} + \left(V_{zu}^2 + 4D_{zzu}R_u\lambda + 4D_{zzu}R_u s\right)^{1/2}}{2D_{zzu}}(z - z_a)\right) \tag{3.228}$$

The gradient at the interface $z = z_a$ is

$$\left.\frac{\partial \bar{c}_u}{\partial z}\right|_{z=z_a} = \frac{V_{zu}\bar{C} + \bar{C}\left(V_{zu}^2 + 4D_{zzu}R_u\lambda + 4D_{zzu}R_u s\right)^{1/2}}{2D_{zzu}} \tag{3.229}$$

Similarly, for the overlying aquitard,

$$\bar{c}_o(x,y,z,s)=\bar{C}\exp\left(\frac{V_{zo}+\left(V_{zo}^2+4D_{zzo}R_o\lambda+4D_{zzo}R_o s\right)^{1/2}}{2D_{zzo}}(z-z_b)\right) \tag{3.230}$$

$$\left.\frac{\partial\bar{c}_o}{\partial z}\right|_{z=z_b}=\frac{V_{zo}\bar{C}+\bar{C}\left(V_{zo}^2+4D_{zzo}R_o\lambda+4D_{zzo}R_o s\right)^{1/2}}{2D_{zzo}} \tag{3.231}$$

Substituting (3.230) and (3.232) into (3.223), we obtain

$$U_m\frac{\partial\bar{C}}{\partial x}-D_{mx}\frac{\partial^2\bar{C}}{\partial x^2}-D_{my}\frac{\partial^2\bar{C}}{\partial y^2}+S_m\bar{C}=0 \tag{3.232}$$

where

$$\begin{aligned}
U_m&=\frac{V_{xm}}{R_m}\\
D_{mx}&=\frac{D_{xxm}}{R_m}\\
D_{my}&=\frac{D_{yym}}{R_m}\\
S_m&=\lambda+s+\gamma_u(\eta_u+s)^{1/2}+\gamma_o(\eta_o+s)^{1/2}\\
\gamma_k&=\frac{n_k(D_{zzk}R_k)^{1/2}}{R_m n_m(z_b-z_a)},\ k=u,o\\
\eta_k&=\frac{V_{zk}^2}{4D_{zzk}R_k}+\lambda=\frac{(V_{zm}n_m)^2}{4D_{zzk}R_k n_k^2}+\lambda,\ k=u,o
\end{aligned} \tag{3.233}$$

The general solution to (3.233) subject to (3.224) in the Laplace domain is

$$\bar{C}(x,y,s)=\sum_{k=0}^{\infty}\left(A_k(s)\exp\left(\frac{U_m+T}{2D_{mx}}\right)+B_k(s)\exp\left(\frac{U_m-T}{2D_{mx}}\right)\right)\cos\frac{\pi ky}{L_2} \tag{3.234}$$

where

$$T = \left(U_m^2 + 4D_{mx}\left(S_m + \beta_k\right)\right)^{1/2}$$

$$\beta_k = D_{my}\left(\frac{\pi k}{L_2}\right)^2$$

$$A_k(s) = \left(\frac{g_k}{s-r_1}\exp(W) - \frac{h_k}{s-r_2}\exp\left(\frac{U_m L_1}{2D_{mx}}\right)\right)\left(\exp(-W) - \exp(W)\right)^{-1}$$

$$B_k(s) = \left(\frac{g_k}{s-r_1}\exp(W) - \frac{h_k}{s-r_2}\exp\left(-\frac{U_m L_1}{2D_{mx}}\right)\right)\left(\exp(W) - \exp(-W)\right)^{-1}$$

$$g_k = \frac{2}{L_2}\int_0^{L_2} C_1(y)\cos\left(\frac{\pi k y}{L_2}\right)dy$$

$$h_k = \frac{2}{L_2}\int_0^{L_2} C_2(g)\cos\left(\frac{\pi k y}{L_2}\right)dy$$

Using the geometric series expansion

$$\left(\exp\left(\frac{L_1 T}{2D_{mx}}\right) - \exp\left(-\frac{L_1 T}{2D_{mx}}\right)\right)^{-1} = \sum_{j=0}^{\infty}\exp\left[-\frac{(1+2j)L_1}{D_{mx}^{1/2}}\left(\frac{U_m^2}{4D_{mn}} + S_m + \beta_k\right)^{1/2}\right] \tag{3.235}$$

we can rewrite A_k and B_k as

$$A_k = \sum_{j=0}^{\infty}\left(\frac{g_k}{s-r_1}\exp\left(\frac{(1+j)L_1 T}{D_{mx}}\right) + \frac{h_k}{s-r_2}\exp\left(-\frac{U_m L_1 + (1+2j)L_1 T}{2D_{mx}}\right)\right)$$

$$B_k = \sum_{j=0}^{\infty}\left(-\frac{g_k}{s-r_1}\exp\left(\frac{jL_1 T}{D_{mx}}\right) - \frac{h_k}{s-r_2}\exp\left(-\frac{U_m L_1 + (1+2j)L_1 T}{2D_{mx}}\right)\right)$$

Therefore, (3.235) can be written as

$$\bar{C}(x,y,s) = \sum_{k=0}^{\infty}\left(M_k + N_k\right)\cos\left(\frac{k\pi y}{L_2}\right) \tag{3.236}$$

where

$$M_k = \sum_{J=0}^{\infty} \frac{g_k}{s-r_1}\left(\exp\left(\frac{xU_m - (x+2jL_1)T}{2D_{mx}}\right) - \exp\left(\frac{xU_m - (x-2(1+2j)L_1)T}{2D_{mx}}\right)\right)$$

$$N_k = \sum_{j=0}^{\infty} \frac{h_k}{s-r_2}\left(\exp\left(\frac{U_m(x-L_1) + (x-(1+2j)L_1)T}{2D_{mx}}\right) - \exp\left(\frac{U_m(x-L_1) - (x+(1+2j)L_1)T}{2D_{mx}}\right)\right)$$

Let

$$\bar{F}_m(\varepsilon_p, r, s) = \frac{1}{s-r}\exp\left[\varepsilon_p\left(\frac{u_m^2}{4D_{mx}} + s_m + \beta_k\right)^{1/2}\right] \tag{3.237}$$

with

$$r = r_1 \text{ or } r_2, \quad p = 1,2,3,4$$

$$\varepsilon_1 = \frac{x+2jL_1}{D_{mx}^{1/2}}, \quad \varepsilon_2 = \frac{x-2(1+j)L_1}{D_{mx}^{1/2}}$$

$$\varepsilon_3 = \frac{x-(1+2j)L_1}{D_{mx}^{1/2}}, \quad \varepsilon_4 = \frac{x+(1+2j)L_1}{D_{mm}^{1/2}}$$

The inverse Laplace transform of (3.238) is

$$F_m(\varepsilon_p, r, t) = \mathscr{L}^{-1}\left[\bar{F}_m(\varepsilon_p, r, s)\right].$$

Finally we get the concentration distribution in the aquifer as

$$C(x,y,t) = \sum_{k=0}^{\infty}(P_k + Q_k)\cos\left(\frac{\pi k y}{L_2}\right) \tag{3.238}$$

where

$$P_k = \sum_{j=0}^{\infty} g_k\left(\exp\left(\frac{xU_m}{2D_{mx}}\right)F_m(\varepsilon_1, r_1, t) - \exp\left(\frac{U_m x}{2D_{mn}}\right)F_m(\varepsilon_2, r_1, t)\right)$$

$$Q_k = \sum_{i=0}^{\infty} h_j\left(\exp\left(\frac{U_m(x-L_1)}{2D_{mx}}\right)F_m(\varepsilon_3, r_2, t) - \exp\left(\frac{U_m(x-L_1)}{2D_{mx}}\right)F_m(\varepsilon_4, r_2, t)\right)$$

The inverse of (3.238) is

$$F_m(\varepsilon_p,r,t)=\frac{2}{\sqrt{\pi}}\int_0^\infty \exp(-\varphi^2-\mu\zeta)\ \mathcal{L}^{-1}\left[\bar{f}_m(\varepsilon_p,r,s)\right]d\varphi,\ \varepsilon_p\neq 0 \tag{3.239}$$

Next,

$$\mathcal{L}^{-1}\left[\bar{f}_m(\varepsilon_p,r,s)\right]=\frac{\mu_u}{4\sqrt{\pi}}\int_0^{t-\mu}P(\tau)Q(\tau)d\tau,\ t>\mu \tag{3.240}$$

with

$$P(\tau)=\exp\left(-\mu_o(\eta_o+r)^{1/2}\right)\operatorname{erfc}(\psi_{o_1})+\exp\left(\mu_o(\eta_o+r)^{1/2}\right)\operatorname{erfc}(\psi_{o_2})$$

$$Q(\tau)=\frac{\exp\left(r\tau-\dfrac{\mu_u^2}{4(t-\mu-\tau)}-\eta_u(t-\mu-\tau)\right)}{(t-\mu-\tau)^{3/2}}$$

$$\psi_{o_1}=\frac{\mu_o}{2\sqrt{T}}-\left[(\mu_o+r)T\right]^{1/2}$$

$$\psi_{o_2}=\frac{\mu_o}{2\sqrt{T}}+\left[(\eta_o+r)\right]^{1/2}$$

Since $t>\mu$, (3.240) reduces to

$$F_m(\varepsilon_p,r,t)=\frac{2}{\sqrt{\pi}}\int_{\varepsilon_k/(2\sqrt{t})}^\infty \exp(-\varphi^2-\mu\zeta)\,\mathcal{L}^{-1}\left[\bar{f}_m(\varepsilon_p,r,s)d\varphi\right],\ \varepsilon_p\neq 0 \tag{3.241}$$

In order to obtain the concentration distribution in the aquitards, we substitute (3.237) into (3.229) and (3.231) and evaluate the inverse Laplace transform as

$$c_u(x,y,z,t)=\sum_{k=0}^{\infty}(T_1(k)+T_2(k))\cos\left(\frac{\pi ky}{L_2}\right) \tag{3.242}$$

with

$$T_1(k) = g_k \exp\left(\frac{V_{zm} n_m (z - z_a)}{2D_{zzu} n_u} + \frac{U_m x}{2D_{mx}}\right) \sum_{j=0}^{\infty} \left(F_u(\varepsilon_1, r_1, t) - F_u(\varepsilon_2, r_1, t)\right)$$

$$T_2(k) = h_k \exp\left(\frac{V_{zm} n_m (z - z_a)}{2D_{zzu} n_u} + \frac{U_m (x - L_1)}{2D_{mx}}\right) \sum_{j=0}^{\infty} \left(F_u(\varepsilon_3, r_2, t) - F_u(\varepsilon_4, r_2, t)\right)$$

where

$$F_u(\varepsilon_p, r, t) = \frac{1}{2} \exp\left[rt + (z - z_a)\left(\frac{R_u r}{D_{zzu}}\right)^{1/2}\right] \operatorname{erfc}\left[(z_a - z)\left(\frac{R_u}{4D_{zzu} t}\right)^{1/2} - (rt)^{1/2}\right]$$
$$+ \frac{1}{2} \exp\left[rt - (z - z_a)\left(\frac{R_u r}{D_{zzu}}\right)^{1/2}\right] \operatorname{erfc}\left[(z_a - z)\left(\frac{R_u}{4D_{zzu} t}\right)^{1/2} + (rt)^{1/2}\right]$$

$$F_u(\varepsilon_p, r, t) = \frac{2}{\sqrt{\pi}} \int_{\varepsilon_p/(2\sqrt{t})}^{\infty} \exp(-\varphi^2 - \mu\zeta) f_u(\varepsilon_p, r, t)\, d\varphi, \quad \varepsilon_p \neq 0;\ p = 1, \ldots, 4$$

$$f_u(\varepsilon_p, r, t) = \frac{\mu_u}{4\sqrt{\pi}} \int_0^{t-\mu} \left\{\exp\left(-\mu_o (\eta_o + r)^{1/2}\right) \operatorname{erfc}(\psi_{o_1}) + \exp\left(-\mu_o (\eta_o + r)^{1/2}\right) \operatorname{erfc}(\psi_{o_2})\right\}$$
$$\times \frac{\exp\left(r\tau - \dfrac{\mu_u^2}{4(t - \mu - \tau)} - \eta_u (t - \mu - \tau)\right)}{(t - \mu - \tau)^{3/2}} d\tau$$

$$\mu_u = \mu\gamma_u - \left(\frac{R_u}{D_{zzu}}\right)^{1/2} (z - z_a)$$

The expression for the concentration distribution in overlying aquitard is same as that for the underlying aquitardwith u replaced by o in the expressions for R, D, F, f and $(z - z_a)$

replaced by $(z - z_b)$.

Example 3.13

This example[43] considers the transport of a sorbing organic contaminant in a converging radial flow field in the vadose zone, and solves the model equations in the Laplace domain.

Let r be the radial coordinate measured from the well center. Then in terms of the following parameters

$$v(r) = -\frac{Q_w}{2\pi b\theta_g r} = \text{ air velocity in the radial direction}$$

$X = r|R_*$ nondimensional length

$$T = \frac{|v(R_*)|t}{R_* R_{m1}} = \frac{Q_w t}{2\pi b\theta_g R_*^2 R_{m1}} = \text{ dimensionless time}$$

$$P = \frac{|v(R_*)|R_*}{D} = \frac{Q_w}{2\pi b\theta_g D} = \text{ Peclet number}$$

the modified dimensionless radially oriented flow equations for a single extraction well can be written as:

$$\frac{\partial C^*}{\partial T} = \frac{1}{P}\frac{\partial^2 C^*}{\partial x^2} + \frac{P+1}{XP}\frac{\partial C^*}{\partial X} - \frac{R_{m2}}{R_{m1}}\frac{\partial S_m^*}{\partial T} - \frac{R_{im1}}{R_{m1}}\frac{\partial C_w^*}{\partial T} - \frac{R_{im2}}{R_{m1}}\frac{\partial S_{im}^*}{\partial T} \tag{3.243}$$

$$\frac{R_{im1}}{PR_{m1}}\frac{\partial C_w^*}{\partial T} + \frac{R_{im2}}{PR_{m1}}\frac{\partial S_{im}^*}{\partial T} = \omega\left(C^* - C_w^*\right) \tag{3.244}$$

$$\frac{R_{m2}}{R_{m1}}\frac{\partial S_m^*}{\partial T} = k_m^o\left(C^* - S_m^*\right) \tag{3.245}$$

$$\frac{R_{im2}}{R_{m1}}\frac{\partial S_{im}^*}{\partial T} = k_{im}^o\left(C^* - S_{im}^*\right) \tag{3.246}$$

where

C^* = normalized contaminant concentration in the gas phase

C_w^* = non-advective domain liquid phase

43 Mark N. Goltz and Mark E. Oxley (1994). An analytical solution to equations describing rate-limited soil vapor extraction of contaminate in the vadose zone. Water Resources Research, Vol. 30, pp. 2691-2698.

S_m^* = normalized solid phase contaminant concentration for the rate-limited sorbent associated with advective domain

S_{im}^* = associated with nonavective domains

ω = Damkohler dimensionless number

K_m^0, K_{im}^0 = Damkohler numbers describing rate limited sorption in the advective and nonavective domains, respectively

D = effective diffusion coefficient of the contaminant in the gas filled pores

$R_{m1}, R_{m2}, R_{im1}, R_{im2}$ = retardation factors

The initial and boundary condition are

$$C^*(X,0) = C_w^*(X,0) = S_m^*(X,0) = S_{im}^*(X,0) = 1,\ X_w < X \le 1 \tag{3.247}$$

$$\left. C^* + \frac{1}{P}\frac{\partial C^*}{\partial X}\right|_{X=1} = 0 \qquad \left.\frac{\partial C^*}{\partial X}\right|_{X=X_w} = 0 \tag{3.248}$$

Applying the Laplace transform to (3.244)-(3.247) and the initial condition (3.248) we obtain

$$s\overline{C^*} - 1 = \frac{d^2\overline{C^*}}{dx^2} + \frac{P+1}{x}\frac{d\overline{C^*}}{dx} - \frac{R_{m2}}{R_{m1}}\left[\overline{S_m^*} - 1\right] - \frac{R_{im1}}{R_{m1}}\left[sS\overline{C_w^*} - 1\right] - \frac{R_{im2}}{R_{m1}}\left[s\overline{S_{im}^*} - 1\right] \tag{3.249}$$

$$\frac{R_{im1}}{PR_{m1}}\left[s\overline{C_w^*} - 1\right] + \frac{R_{im2}}{PR_{m1}}\left[s\overline{S_{im}^*} - 1\right] = \omega\left(\overline{C^*} - \overline{C_w^*}\right) \tag{3.250}$$

$$\frac{R_{m2}}{PR_{m1}}\left[s\overline{S_m^*} - 1\right] = k_m^0\left(\overline{C^*} - \overline{S_m^*}\right) \tag{3.251}$$

$$\frac{R_{im2}}{PR_{m1}}\left[s\overline{S_{im}^*} - 1\right] = k_{im}^0\left(\overline{C^*} - \overline{S_{im}^*}\right) \tag{3.252}$$

The boundary conditions (3.249) transform to

$$\left.\overline{C^*} + \frac{1}{P}\frac{d\overline{C^*}}{dX}\right|_{X=1} = 0$$
$$\left.\frac{d\overline{C^*}}{dX}\right|_{X=X_w} = 0 \qquad (3.253)$$

Using equations (3.250)-(3.253), we get the differential equation

$$\frac{d^2\overline{C^*}}{dX^2} + \frac{P+1}{X}\frac{d\overline{C^*}}{dX} - \gamma\overline{C^*} = -\beta \qquad (3.254)$$

where $\gamma = s\beta$ and

$$\beta = 1 + k_m^o P R_{m2}\left(sR_{m2} + k_m^o P R_m\right)^{-1} + \omega P\left[R_{im1}R_{im2}s + k_{im}^o R_{m1}\left(R_{im1} + R_{im2}\right)P\right]G(s)$$

with

$$G(s) = \left\{R_{im1}R_{im2}s^2 + \left[k_{im}^2\left(R_{im1} + R_{im2}\right) + \omega R_{im2}\right]\cdot PR_{m1}s + \omega k_{im}^o P^2 R_{m1}^2\right\}^{-1}$$

The general solution of (3.255) is

$$\overline{C^*}(X,s) = \frac{1}{s} + AX^{-\nu}I_\nu(qX) + BX^{-\nu}I_{-\nu}(qX) \qquad (3.255)$$

where

$q = \gamma^{0.5}$
$\nu = P/2$
$A, B =$ coefficients dependent on s
$I_\nu =$ a modified Bessel function

Applying the boundary conditions (3.254) to (3.256), we obtain

$$A = \frac{P}{s}\frac{G_{-\nu}}{G_\nu H_{-\nu} - G_{-\nu}H_\nu}$$
$$B = -\frac{P}{s}\frac{G_\nu}{G_\nu H_{-\nu} - G_{-\nu}H_\nu}$$

where

$$G_{\nu} = -\nu I_{\nu}(qX_w) + qX_w I'_{\nu}(qx_w)$$
$$G_{-\nu} = -\nu I_{\nu}(qX_w) + qX_w I'_{-\nu}(qx_w)$$
$$H_{\nu} = \nu I_{\nu}(q) + qI'_{\nu}(q)$$
$$H_{-\nu} = \nu I_{\nu}(q) + qI'_{-\nu}(q)$$

depend on s. Substituting these into (3.256) we obtain

$$\overline{C^*}(X,s) = \frac{1}{s} - \frac{PX^{-\nu}}{s}\left[\frac{G_{\nu}I_{-\nu}(qX) - G_{-\nu}I_{\nu}(qX)}{G_{\nu}H_{-\nu} - G_{-\nu}H_{\nu}}\right] \tag{3.256}$$

Using recurrence relations for I_{ν}, we obtain

$$G_{\nu} = qX_w I_{\nu+1}(qX_w)$$
$$G_{-\nu} = qX_w I_{-\nu-1}(qX_w)$$
$$H_{\nu} = qI_{\nu-1}(q)$$
$$H_{-\nu} = qI_{-\nu+1}(q)$$

The solution of (3.255) subject to (3.254) is

$$\overline{C^*}(X,s) = \frac{1}{s} - \frac{PX^{-\nu}}{sq}\left[\frac{I_{\nu+1}(qX_w)I_{-\nu}(qX) - I_{-\nu-1}(qX_w)I_{\nu}(qX)}{I_{\nu+1}(qX_w)I_{-\nu+1}(q) - I_{-\nu-1}(qX_w)I_{\nu-1}(q)}\right] \tag{3.257}$$

The Wronskian is evaluated as

$$W[I_{\nu}, I_{-\nu}](qx_w) = I_{\nu}(qX_w)I'_{-\nu}(qX_w) - I'_{\nu}(qX_w)I_{-\nu}(qX_w) = -\frac{2\sin(\nu\pi)}{\pi qX_w}.$$

At the extraction well $(X = X_w)$, (3.258) simplifies to

$$\overline{C^*}(X_w,s) = \frac{1}{s} - \frac{1}{s^2}\frac{2PX_w^{-\nu-1}\sin(\nu\pi)}{\pi\beta\left[I_{\nu+1}(qX_w)I_{-\nu+1}(q) - I_{-\nu-1}(qX_w)I_{\nu-1}(q)\right]} \tag{3.258}$$

Next, the total normalized dimensionless mass is

$$M(T) = \frac{2\int_{X_w}^{1}\left(R_{m1}C^* + R_{m2}S_m^* + R_{im1}C_w^* + R_{im2}S_{im}^*\right)XdX}{\left(R_{m1} + R_{m2} + R_{im1} + R_{im2}\right)\left(1 - X_w^2\right)} \tag{3.259}$$

The Laplace transform of (3.260) is

$$\overline{M}(s)=\frac{2\int_{X_w}^{1}\left(R_{m1}\overline{C^*}+R_{m2}\overline{S_m^*}+R_{im1}\overline{C_w^*}+R_{im2}\overline{S_{im}^*}\right)XdX}{\left(R_{m1}+R_{m2}+R_{im1}+R_{im2}\right)\left(1-X_w^2\right)} \tag{3.260}$$

Using (3.250) – (3.253),(3.261) simplifies to

$$\overline{M}(s)=\frac{2\cdot\int_{X_w}^{1}\left(D\overline{C^*}+E\right)XdX}{\left(R_{m1}+R_{m2}+R_{im1}+R_{im2}\right)\left(1-X_w^2\right)} \tag{3.261}$$

where the constants *D* and *E* are given by

$$D=R_{m1}\beta$$
$$E=\left(\left(1-\beta\right)R_{m1}+R_{m2}+R_{im1}+R_{im2}\right)/s$$

Solution of (3.255) subject to (3.254) is

$$\int_{X_w}^{1}\overline{C^*}(X)dX=-\frac{P}{\gamma}\overline{C^*}(X_w,s)+\frac{\left(1-X_w^2\right)}{2s} \tag{3.262}$$

It follows that

$$\overline{M}(s)=\frac{1}{s}-\frac{2PR_{m1}\overline{C^*}(X_w,s)}{s\left(R_{m1}+R_{m2}+R_{im1}+R_{im2}\right)\left(1-X_w^2\right)} \tag{3.263}$$

Substituting for $\overline{C^*}(X_w,s)$ from (3.259), we obtain the Laplace domain solution

$$\overline{M}(s)=\frac{1}{s}-\frac{U}{s^2}+\frac{V}{s^3\left[I_{\nu+1}(qX_w)I_{-\nu+1}(q)-I_{-\nu-1}(qX_w)I_{\nu-1}(q)\right]} \tag{3.264}$$

where *U* and *V* are independent of *s* and are given below:

$$U=\frac{2PR_{m1}}{\left(R_{m1}+R_{m2}+R_{im1}+R_{im2}\right)\left(1-X_w^2\right)}$$
$$V=\frac{4P^2R_{m1}X_w^{-\nu-1}\sin(\nu\pi)}{\pi\beta\left(R_{m1}+R_{m2}+R_{im1}+R_{im2}\right)\left(1-X_w^2\right)}.$$

Example 3.14

In this example[44], the Laplace-transform analytic element method (LT-AEM)[45],[46], is used to solve the 2D transient confined flow problem governed by linear diffusion. The mathematical flow model for this situation is given by

$$\frac{\partial^2 \Phi}{\partial x^2} + \frac{\partial^2 \Phi}{\partial y^2} = \frac{1}{\alpha} \frac{\partial \Phi}{\partial t} \tag{3.265}$$

where

$K =$ hydraulic conductivity

$h =$ hydraulic head

$S_S =$ specific storage

$\Phi = Kh =$ discharge potential

$\alpha = K / S_S$

Applying the Dupuit assumption (van de Giesen et al.[47], 1994), the linearized Laplace transform of the above equation is

$$\frac{\partial^2 \bar{\Phi}}{\partial x^2} + \frac{\partial^2 \bar{\Phi}}{\partial y^2} = \frac{1}{\alpha}\left[p\bar{\Phi} - \Phi_0 \right] \tag{3.266}$$

where

$\bar{\Phi}$ = the Laplace transform of Φ

44 Alex Furman and Shlomo P. Neuman(2003). Laplace-transform analytic element solution of transient flow in porous media. Advances in Water Resources Vol. 26, pp. 1229–1237

45 F. R. de Hoog, J. H. Knight, and A. N. Stokes (1982). An improved method for numerical inversion of Laplace transforms. SIAM Journal of Statistical Computing, 3(3) : 357-366.

46 K. Kuhlman(2008). Laplace Transform Analytic Element Method: A Semi-analytic Solution for Transient GroundwaterFlow Simulation. VDM Verlag.

47 N. C. van de Giesen, J.-Y. Parlange, and T. S. Steenhuis (1994). Transient flow to open drains: Comparison of linearized solutions with and without the Dupuit assumption. Water Resources Research, Vol. 30, pp. 3033-3039.

In the general case (initial discharge potential $\Phi_0 \neq 0$, the solution in the Laplace domain is given by

$$\mathcal{L} = \mathcal{L}_1 + \mathcal{L}_2$$

$\mathcal{L}_1$ = the Laplace domain solution of the modified Helmholtz equation (when $\Phi_0 \equiv 0$)

$\mathcal{L}_2$ = the Laplace domain solution of a Poisson equation with a nonzero source term Φ_0.

In the case of an instantaneous point source of strength q at the origin, the discharge potential is given by

$$\Phi = \frac{q}{4\pi\alpha(t-\tau)} \exp\left(-\frac{r^2}{4\alpha(t-\tau)}\right) \tag{3.267}$$

where

$q =$ the volume of water injected per unit specific storage $= V / S_S$

$r =$ the radial distance from the origin

which leads to

$$d\Phi = \frac{c(\tau)d\tau}{4\pi\alpha(t-\tau)} \exp\left(-\frac{r^2}{4\alpha(t-\tau)}\right) \tag{3.268}$$

where

$$c(\tau) = \frac{Q(\tau)}{S_S} = \text{a finite time rate} = \frac{dq}{d\tau}$$

with Q being the volumetric rate of water injection.

Integrating the above equation, we obtain

$$\Phi = \frac{1}{4\pi} \int_0^t \frac{Q(\tau)}{(t-\tau)} \exp\left(-\frac{r^2}{4\alpha(t-\tau)}\right) d\tau \tag{3.269}$$

which has the Laplace transform

$$\bar{\Phi} = \frac{\bar{Q}(\tau)}{2\pi} K_0\left(r\sqrt{\frac{p}{\alpha}}\right) \tag{3.270}$$

where K_0 is the modified Bessel function of the second kind of order 0. This turns out to be a solution of the modified Helmholtz equation in radial coordinates,

$$\frac{\partial^2 \bar{\Phi}}{\partial r^2} + \frac{1}{r}\frac{\partial \bar{\Phi}}{\partial r} - \frac{p}{\alpha}\bar{\Phi} = 0. \tag{3.271}$$

The Laplace transform $\bar{\Phi}$ above satisfies the flow equation in the Laplace domain when $\Phi_0 \equiv 0$, and hence provides us the Laplace domain solution $\mathscr{L}_1$ mentioned earlier.

In the case when injection occurs at a constant rate Q_0 starting at time τ,

$$Q(t|\tau) = Q_0 H(t-\tau) \tag{3.272}$$

where H is the Heaviside function. Since

$$\bar{Q}(p|\tau) = Q_0 \exp(-pt)/p,$$

equation (3.271) becomes

$$\bar{\Phi} = \frac{Q_0 \exp(-pt)}{2\pi p} K_0\left(r\sqrt{\frac{p}{\alpha}}\right). \tag{3.273}$$

In the case of circular inhomogeneity, using separation of variables on the modified Helmholtz equation, we obtain

$$\bar{\Phi} = \sum_{n=0}^{\infty} R_n(r)\Psi_n(\psi) \tag{3.274}$$

where

$r =$ the radial distance from the center of he inhomogeneity

$\psi =$ the polar angle from the positive x-axis, going counter-clockwise

which has the solution

$$\begin{aligned} R_n(r) &= A_n J_n\left(i\sqrt{\frac{p}{\alpha}}r\right) + B_n Y_n\left(i\sqrt{\frac{p}{\alpha}}r\right) \\ \Psi_n(\psi) &= C_n \sin(n\psi) + D_n \cos(n\psi) \end{aligned} \tag{3.275}$$

where

A_n, B_n, C_n, D_n = constant coefficients

J_n = Bessel function of first kind of order n

Y_n = Bessel function of second kind of order n.

Elimination of the complex argument i yields the Laplace domain solution in terms of the modified Bessel functions:

$$\bar{\Phi} = \sum_{n=0}^{\infty}\left[a_n I_n(P)\sin(n\psi) + b_n I_n(P)\cos(n\psi) + c_n K_n(P)\sin(n\psi) + d_n K_n(P)\cos(n\psi)\right] \quad (3.276)$$

where

$P = i\sqrt{\frac{p}{\alpha}}r$

a_n, b_n, c_n, and d_n are constant coefficients, and

I_n = modified Bessel function of first kind of order n

K_n = modified Bessel function of second kind of order n.

Numerical methods can be used to invert the Laplace domain solution.

Example 3.15

In this example[48]the movement of colloid concentrations in aqueous and solid phases is considered, and analytical solutions are obtained using the Laplace transform and matrix decomposition. The general equation for colloid transport governed by advection, dispersion, and exchange between the liquid phase of the low (1) and high (2) permeability regions is given by

$$\rho\frac{\partial S_{ai}}{\partial t}+\rho\frac{\partial S_{si}}{\partial t}+\theta_i\frac{\partial C_i}{\partial t}=\theta_i D_i\frac{\partial^2 C_i}{\partial x^2}-\theta_i v_i\frac{\partial C_i}{\partial x}+\alpha(C_j-C_i)\ (i=1,2;j=1,2) \tag{3.277}$$

where

t = time[T]

x = distance [L],

Di = longitudinal dispersion coefficient $[L^2/T]$,

v_i = pore-water velocity [L/T] with $v_1 < v_2$

C_i = the liquid phase concentration of region i $[Nc\ L^{-3}]$

S_{ai} = the solid phase concentrations for reversibly retained colloids $[N_c\ M^{-1}]$

S_{si} = the solid phase concentrations for irreversibly retained colloids $[N_c\ M^{-1}]$

θ_i = volume of water in region i per bulk volume $[L^3{}_{w,i}\ L^{-3}]$

ρ = soil bulk density assumed to be equal for both soils $[ML^{-3}]$, and

α = the coefficient for mass transfer between the two aqueous regions $[T^{-1}]$.

There are two kinetic retention sites, with the first site employing a conventional attachment and detachment model to describe behavior of reversibly retained colloids to and from the aqueous and solid phases, and the second one governed by irreversible colloid retention due to straining or attachment:

$$\text{Site 1:}\quad \rho\frac{\partial S_{ai}}{\partial t}=\theta_i k_{ai}C_i-\rho k_d S_{ai},\ i=1,2 \tag{3.278}$$

$$\text{Site 2:}\quad \rho\frac{\partial S_{si}}{\partial t}=\theta_i k_{ai}C_i,\ i=1,2 \tag{3.279}$$

---

48 Feike J. Leij and Scott A. Bradford (2013). Colloid transport in dual-permeability media. Journal of Contaminant Hydrology, Vol. 150, pp. 65–76

where

k_{ai} = the coefficient for attachment of colloids from aqueous region *i* onto the solid phase $[T^{-1}]$

k_d = the coefficient for detachment of colloids from the solid into the aqueous phase $[T^{-1}]$

k_{si} = the coefficient for irreversible colloid retention from the aqueous region *i* into the solid phase $[T^{-1}]$

Consider the case of a medium initially free of colloids with a step pulse C_0 applied at the inlet and a zero-gradient outlet concentration at distance =,

$$C_i - \delta\kappa \frac{\partial C_i}{\partial x} = C_0$$
$$\delta = \begin{cases} 0 \text{ for first type inlet condition, i.e., flux-averaged detection mode} \\ 1 \text{ for third type inlet condition, i.e., volume-averaged detection mode} \end{cases} \tag{3.280}$$

$$C_i(x,0) = 0 \tag{3.281}$$

$$\frac{\partial C_i}{\partial x}(\infty, t) = 0$$

where

κ = dispersivity [L]

C_0 = concentration of the applied solution

Applying the Laplace transform to the flow equation, the two rate equations, and the boundary conditions, we get

$$\kappa v_i \frac{d^2\overline{C}_i}{dx^2} - v_i \frac{d\overline{C}_i}{dx} = \left[s\left(1 + \frac{k_{ai}}{s + k_d}\right) + k_{si} + \alpha_i \right]\overline{C}_i - \alpha_i \overline{C}_j, \ (i = 1,2;\ j = 1,2) \tag{3.283}$$

The solution for the first region is:

$$\overline{C}_1(x,s) = \frac{C_0}{2}\exp\left(\frac{x}{2\kappa}\right) \times \{A_1 \exp(-A_2 x) - A_3 \exp(-A_4 x)\} \tag{3.284}$$

where

$$A_1 = \left[\frac{a_1}{r} + \frac{1}{s} - \frac{1}{rs}\left(\frac{b_1}{s+k_d} - d_4\right)\right]$$

$$A_2 = \left(\frac{a_2 s - \dfrac{b_2}{s+k_d} + d_2 + \dfrac{\nu_1\nu_2}{2\kappa} + r}{2\nu_1\nu_2\kappa}\right)^{1/2}$$

$$A_3 = \left[\frac{a_1}{r} - \frac{1}{s} - \frac{1}{rs}\left(\frac{b_1}{s+k_d} - d_4\right)\right]$$

$$A_4 = \left(\frac{a_2 s - \dfrac{b_2}{s+k_d} + d_2 + \dfrac{\nu_1\nu_2}{2\kappa} - r}{2\nu_1\nu_2\kappa}\right)^{1/2}$$

$$a_1 = \nu_2 - \nu_1$$

$$a_2 = \nu_2 + \nu_1$$

$$b_1 = (k_{a1}\nu_2 - k_{a2}\nu_1)/k_d$$

$$b_2 = (k_{a1}\nu_2 + k_{a2}\nu_1)/k_d$$

$$d_1 = (k_{a1} + k_{s1} + \alpha_1)/\nu_2 - (k_{a2} + k_{s2} + \alpha_2)/\nu_1$$

$$d_2 = (k_{a1} + k_{s1} + \alpha_1)/\nu_2 + (k_{a2} + k_{s2} + \alpha_2)/\nu_1$$

$$d_3 = (k_{a1} + k_{s1} + \alpha_1)/\nu_2 - (k_{a2} + k_{s2} - \alpha_2)/\nu_1$$

$$d_4 = (k_{a1} + k_{s1} - \alpha_1)/\nu_2 + (k_{a2} + k_{s2} + \alpha_2)/\nu_1.$$

Example 3.16

The problem of the estimation of backward spreading of the initial plume when tracers are into an injection borehole is the topic of this example[49]. The method of Green's function is used to find the Laplace domain solution for solute concentration in an infinite porous medium. Breakthrough curves are obtained by numerically inverting the Laplace transform.

Assuming that (i) a tracer is introduced without disturbing the velocity in the aquifer,

(ii) that mechanical dispersion is governed by Fick's law, and (iii) the dispersion coefficient is proportional to the velocity, the mathematical model describing the flow is

$$\frac{a_L A}{r}\frac{\partial^2 C}{\partial r^2}+\frac{A}{r}\frac{\partial C}{\partial r}+\frac{a_L A}{r}\frac{\partial^2 C}{\partial r^2}+\frac{M}{2\pi R b n_e}\delta(R-r)=\frac{\partial C}{\partial t} \tag{3.285}$$

where

$C=$ the concentration

$a_L=$ the longitudinal dispersivity

$R=$ the radial distance between the pumping and injection wells

$M=$ the tracer mass

$A=\dfrac{Q}{2\pi b n_e}$

$b=$ aquifer thickness

$n_e=$effective porosity

The initial and boundary conditions are:

$$\begin{aligned}
&C(r,0)=0,\ r_c\le r<\infty \text{ (the initial tracer concentration = 0)}\\
&\frac{\partial C(r_c,t)}{\partial r}=0,\ t>0 \text{ (zero concentration gradient at the well-aquifer interface)}\\
&\lim_{r\to\infty} C(r,t)=0
\end{aligned} \tag{3.286}$$

49 Jui-Sheng Chen, Chew-Wuing Liu, Chia-Shyun Chen, Hun-Der Yeh(1996). A Laplace transform solution for tracer tests in a radially convergent flow field with upstream dispersion. Journal of Hydrology Vol. 183, pp. 263-275

The flow model and the boundary conditions can be expressed in terms of dimensionless time τ and dimensionless radial distance ρ as

$$\frac{1}{\rho}\frac{\partial^2 C}{\partial \rho^2}+\frac{1}{\rho}\frac{\partial C}{\partial \rho}+C_r\delta(\rho-\rho_o)\delta(\tau)=\frac{\partial C}{\partial \tau} \tag{3.287}$$

$$C(\rho,0)=0,\ \rho_C\le\rho<\infty$$
$$\frac{\partial C(\rho_C,\tau)}{\partial \rho}=0,\ \tau>0$$
$$\lim_{\rho\to\infty} C(\rho,\tau)=0 \tag{3.288}$$

where

$$\rho=\frac{r}{a_L}$$
$$\rho_o=\frac{R}{a_L}$$
$$\tau=\frac{At}{L^2}$$
C_r = the reference concentration.

Taking the Laplace transform of the flow equation (3.288) with respect to τ:

$$\frac{1}{\rho}\frac{d^2\bar{C}}{d\rho^2}+\frac{1}{\rho}\frac{d\bar{C}}{d\rho}-p\bar{C}=-C_r(\rho-\rho_o) \tag{3.289}$$

$$\frac{d\bar{C}(\rho,p)}{d\rho}=0 \text{ at } \rho_C$$
$$\lim_{\rho\to\infty}\bar{C}(\rho,p)=0 \tag{3.290}$$
$$\bar{C}(\rho,p)=\int_0^\infty C(\rho,\tau)e^{-p\tau}d\tau$$

The above being an inhomogeneous equation, its complete solution consists of a homogeneous solution and a particular solution; this is achieved by converting it into its self-adjoint form and then applying Green's function.

The self-adjoint form of the model is

$$\frac{d^2\overline{C}_1}{d\rho^2}-\left(p\rho+\frac{1}{4}\right)\overline{C}_1=-F(\rho) \tag{3.291}$$

where

$$\overline{C}_1=e^{\rho/2}\overline{C}$$
$$F(\rho)=\rho e^{\rho/2}C_r\delta(\rho-\rho_0)$$

and the boundary conditions are

$$\left.\frac{d\overline{C}_1}{d\rho}-\frac{1}{2}\overline{C}_1\right|_{\rho=\rho_c}=0$$
$$\overline{C}_1\sim O(e^{-\rho/2})\text{ as }\rho\to\infty \tag{3.292}$$

This is a regular Sturm-Liouville equation[50], with the general solution given by

$$\overline{C}_1(\rho,p)=\int_{\rho_c}^{\infty}g(\rho,p,\xi)F(\xi)d\xi \tag{3.293}$$

where $g(\rho,p,\xi)$ is the Green's function for the above problem given by[51]:

$$g(\rho,p,\xi)=\begin{cases}g_1=(\pi/p^{1/3})Ai(s)[Bi(z)-X.Ai(z)] & \text{if } \rho_c\le\rho<\xi\\ g_2=(\pi/p^{1/3})Ai(z)[Bi(s)-X.Ai(s)] & \text{if } \xi\le\rho<\infty\end{cases}$$

$$s=p^{1/3}\left(\xi+\frac{1}{4p}\right)$$

$$z=p^{1/3}\left(\rho+\frac{1}{4p}\right)$$

$$X=\frac{2p^{1/3}Bi'(z_c)-Bi(z_c)}{2p^{1/3}Ai'(z_c)-Ai(z_c)}$$

---

50 W.N. Everitt (2005). A catalogue of Sturm–Liouville differential equations. Sturm–Liouville Theory, Past and Present (2005), pp. 271–331.

51 C.S. Chen and G. D. Woodside(1988). Analytical Solution for Aquifer Decontamination by Pumping. Water Resources Research. Vol. 24, pp 1329-1338.

Example 3.17

Analytical porosity-weighted solutions of the one-dimensional dual-porosity model are derived in this example.[52] The groundwater diffusion equation describing flow is

$$K_c \frac{\partial^2 H_c}{\partial x^2} = S_c \frac{\partial H_c}{\partial t} + S_m \frac{\partial H_m}{\partial t} \tag{3.294}$$

where

$K_c =$ hydraulic conductivity of the high-permeability continuum (channel)
$S_c =$ storage cefficient of the high-permeability continuum (channel)
$S_m =$ storage cefficient of the low-permeability continuum (matrix)
$H_c = H_c(x,t) =$ hydraulic head of the channel
$H_m = H_m(x,t) =$ hydraulic head of the matrix
$x =$ distance
$t =$ time

The term $S_m \frac{\partial H_m}{\partial t}$ on the right hand side of the above flow model represents the source term from the matrix.

Letting $\varphi = V_c/V_a$, $1 - \varphi = V_c/V_{a,}$ where V_c = the channel volume, V_m = the matrix volume, and V_c= the apparent total volume, and using φ and $1-\varphi$ as weights, we obtain

$$\phi K_c \frac{\partial^2 H_c}{\partial x^2} = \phi S_c \frac{\partial H_c}{\partial t} + (1-\phi) S_m \frac{\partial H_m}{\partial t} \tag{3.295}$$

or equivalently

$$K_c \frac{\partial^2 H_c}{\partial x^2} = S_c \frac{\partial H_c}{\partial t} + \beta S_m \frac{\partial H_m}{\partial t} \tag{3.296}$$

with

$$\beta = \frac{(1-\phi)}{\phi}.$$

---

[52] F. Cornaton and P. Perrochet (2002). Analytical 1D dual-porosity equivalent solutions to 3D discrete single-continuum models. Applications to karstic spring hydrograph modelling. Journal of Hydrology 262 (2002) 165-176.

In the special case of both channel and matrix being tubular,

$$\beta = \left(\frac{r_m}{r_c}\right)^2$$

r_m = radius of the matrix

r_m = radius of the channel.

The boundary conditions are

$$H_c(0,t) = BC(t) \text{ or } -K_c \frac{\partial H_c}{\partial x}(0,t) = BC(t)$$
$$H_c(\infty,t) = 0 \tag{3.297}$$

where $BC(t)$ is a transient input at $x = 0$.

Assuming that exchanges between porosities are of first order, the matrix storage term in the flow model can be written as

$$S_m \frac{\partial H_m}{\partial t} = -\alpha\left[H_m(x,t) - H_c(x,t)\right] \tag{3.298}$$

Applying the Laplace transform to the flow model, the boundary conditions, and the initial conditions $H_m(x,0) = H_c(x,0) = 0$, we get

$$pS_c\overline{H}_c + p\beta S_m\overline{H}_m = K_c \frac{\partial^2 \overline{H}_c}{\partial x^2}$$
$$\overline{H}_m = \frac{\overline{H}_c}{1 + \frac{pS_m}{\alpha}} \tag{3.299}$$

where $\overline{H}_i = \mathcal{L}(H_i)$ is the Laplace transform of H_i.

Further simplification yields

$$\frac{\partial^2 \bar{H}_c}{\partial x^2} = A(p)\bar{H}_c \text{ with } A(p) = \frac{p}{K_c}\left(S_c + \frac{\beta S_m}{1 + \dfrac{pS_m}{\alpha}}\right)$$
$$\bar{H}_c(0,p) = \mathcal{L}(BC)(p) \text{ or } -K_c\frac{\partial \bar{H}_c}{\partial x}(0,p) = \mathcal{L}(BC)(p) \tag{3.300}$$
$$\bar{H}_c(\infty,p) = 0$$

where $\mathcal{L}(BC)(p)$ is the Laplace transform of the boundary condition $BC(t)$ at $x = 0$.

The above equation has the general solution

$$\bar{H}_c(x,p) = \frac{1}{2}\left(\left[\bar{F}_1(p) + \bar{F}_2(p)\right]e^{x\sqrt{A(p)}} + \left[\bar{F}_2(p) - \bar{F}_1(p)\right]e^{-x\sqrt{A(p)}}\right) \tag{3.301}$$

where $\bar{F}_1(p)$ and $\bar{F}_2(p)$ are the Laplace transforms of the boundary conditions.

Using the boundary condition $\bar{H}_c(0,p) = 0$, we obtain

$$\bar{H}_c(x,p) = \mathcal{L}(BC)(p)e^{-x\sqrt{A(p)}}. \tag{3.302}$$

The solutions for the other boundary conditions are similarly obtained.

Example 3.18

In this section, a few applications of the Laplace transform from finance. In the first application, the present value $V(t)$in finance[53]for a given cash flow $C(t)$ with a specified rate of discount $r$is derived. The present value function for cash flow $C(t)$ is calculated from the following equation:

$$V(r) = \int_0^\infty e^{-rt} C(t)dt \tag{3.303}$$

In other words, the present value is simply the Laplace transform of the cash flow, $C(t)$. Using this fact, the following table of present value - cash flow pairs is easily obtained. Most of the formulas shown in the above table can be derived from the time-derivative property of the Laplace transform (Theorem 2.3): $\mathcal{L}\{C'\} = s\mathcal{L}\{C\} - C(0)$.

Cash Flow $C(t)$	Present Value $V(r)$
1	$1/r$
e^{at}	$1/(r-a),\ a<r$
t	$1/r^2$
t^n	$n!/r^n$
$\sin(t)$	$1/(r^2+1)$
$\cos(t)$	$r/(r^2+1)$
$e^{at}C(t)$	$V(r-a),\ a<r$
$C'(t)$	$rV(r)-C(0)$
$\int_0^t C(x)dx$	$V(r)/r$

53 Stephen A. Buser (1986). Laplace Transforms as Present Value Rules:A Note. The Journal of Finance, pp. 243-247.

The above formula can be rewritten as

$$\mathcal{L}\{C\} = \left(\mathcal{L}\{C'\} + C(0)\right)/r.$$

For example if $C(t) = e^{at}$, then

$$C(0) = 1,\ C'(t) = ae^{at},\ \text{and } \mathcal{L}(e^{at}) = \left(1 + a\mathcal{L}(e^{at})\right)/r$$

which simplifies to

$$\mathcal{L}(e^{at}) = 1/(r-a).$$

In the second application, the Black-Scholes equation for calculating option prices is solved by the use of Laplace transform[54]. The European option price *u(x,t)* for underlying asset price x, strike price K, time to maturity t, and expiration date of the option *T* is given by the Black-Scholes equation

$$\frac{\partial u}{\partial t} - \frac{1}{2}\sigma^2 x^2 \frac{\partial^2 u}{\partial x^2} - rx\frac{\partial u}{\partial x} + ru = 0;\ 0 < x < \infty,\ 0 < t \le T \tag{3.304}$$

where

σ = the volatility of the underlying asset

r = the risk-free interest rate of the market

In the case of *n* assets $\mathbf{x} = (x_1, x_2, \ldots, x_n)$ the Black-Scholes equation becomes

$$\frac{\partial u}{\partial t} - \frac{1}{2}\sum_{i=1}^{n}\sum_{j=1}^{n} a_i a_j x_i x_j \frac{\partial^2 u}{\partial x_i \partial x_j} - \sum_{i=1}^{n} r x_i \frac{\partial u}{\partial x} + ru = 0,\ 0 < x_i < \infty,\ i = 1, 2, \ldots, n \tag{3.305}$$

where

$$a_{ij} = \sum_{k=1}^{n} \sigma_{ik}\sigma_{jk}$$

σ_{ij} = correlation between the assets x_i and x_j.

Equation (3.304) can be solved numerically by using the time-discretization method[55].

---

54 Hyoseop Lee and Dongwoo Sheen (2009). Laplace transformation method for the Black-Scholes equation. International journal of numerical analysis and modeling, Volume 6, Number 4, pp. 642-658.

55 W. McLean, I. H. Sloan, and V. Thomee. Time discretization via Laplace transformation of an integro-differential equation of parabolic type. Numer. Math., 102:497{522, 2006.

Example 3.19

In this example[56], the method of moment generating function (mgf) is used to obtain the probability distribution of sums of independent discrete random variables. The mgfof a random variable X is defined[57] as the expected value of the function e^{tX}:

$$M_X(t) = E\left[e^{tX}\right] = \mathcal{L}(-t) \text{ for some real number } t \tag{3.306}$$

The mgf is computed by the formula:

$$M_X(t) = \begin{cases} \sum_{\text{all } x} e^{tx} f(x), & X \text{ is discrete} \\ \int_{-\infty}^{\infty} e^{tx} f(x)dx, & X \text{ is continuous} \end{cases} \tag{3.307}$$

The mgf of a probability distribution is unique, *i.e.*, there exists a one-to-one correspondence between the probability distribution and its mgf. Suppose a discrete rvX has the probability distribution:

$$f(x_j) = P(X = x_j) = p_j,\ j = 1,\ 2, \dots,\ k.$$

Then its mgf is:

$$M_X(t) = p_1 e^{tx_1} + p_2 e^{tx_2} + \dots + p_k e^{tx_k}.$$

In other words, if the rvX has the mgf:

$$M_X(t) = p_1 e^{tx_1} + p_2 e^{tx_2} + \dots + p_k e^{tx_k},$$

then its probability distribution is:

$$f(x_j) = P(X = x_j) = p_j,\ j = 1,\ 2, \dots,\ k. \tag{3.308}$$

---

56 Rohan J. Dalpatadu, and Ashok K. Singh (2008). A numerical method for computing the probability distribution of a finite sum of independent nonnegative random variables. Advances and Applications in Statistics, Vol. 9, No. 1, pp. 145-152.

57 R. L. Scheaffer & L. J. Young (2009). Introduction to Probability and Its Applications, Third Edition. Cengage Learning.

The mgf, as the name suggests, can be used to compute the moments (such as the mean and variance) in one of two ways:

(a) The k-th moment $E\left[X^k\right]$ can be obtained by finding the *k-th* derivative of the mgf with respect to t at the origin $t=0$:

$$E\left[X^k\right]=\left.\frac{d^k M_X(t)}{dt^k}\right|_{t=0}$$

The k-th moment $E\left[X^k\right]$ can be obtained by expanding the mgf as a Taylor series and finding the coefficient of the term $\frac{t^k}{k!}$:

$$M_X(t)=1+tE[X]+\frac{t^2}{2!}E\left[X^2\right]+\frac{t^3}{3!}E\left[X^3\right]+\ldots+\frac{t^k}{k!}E\left[X^k\right]+\ldots$$

$$E\left[X^k\right]=\text{ coefficient of }\frac{t^k}{k!}\text{ in the series expansion.}$$

Once the moments $E[X]$ and $E\left[X^2\right]$ have been calculated, the variance of the rv can be computed by the well-known formula:

$$Var(X)=E\left[X^2\right]-\{E[X]\}^2.$$

We next show how the method of mgf can be used to compute the probability distribution of the sum of *k* dice (*k*≥ 2). Consider the experiment of rolling *k* fair dice, and let X_i represents the number that comes up when *i-th*fair die is rolled, $i=1,\ 2,\cdots,\ k.$ In this paper, we derive the probability distribution of the sum X. The probability distribution of each X_i is given by:

$$f(x)=\begin{cases}\frac{1}{6} & x=1,\ 2,\cdots,\ 6\\ 0 & \text{otherwise}\end{cases}$$

and its moment generating function (*mgf*) is:

$$M_{X_i}(t)=E(e^{tX_i})=\frac{1}{6}\left(e^t+e^{2t}+e^{3t}+\cdots e^{6t}\right).$$

Since the random variables $X_1, X_2, \ldots, X_k$ are independent, the *mgf* of the sum S is:

$$M_S(t) = E\left[e^{tX}\right] = E\left[e^{t(X_1+\cdots+X_k)}\right] = \prod_{i=1}^{k} E\left[e^{tX_i}\right]$$
$$= \prod_{i=1}^{k}\left[\frac{1}{6}\left(e^{t} + e^{2t} + \cdots + e^{6t}\right)\right] = \frac{1}{6^k}\left(e^{t} + e^{2t} + \cdots + e^{6t}\right)^k. \quad (3.309)$$

We can now expand the right hand side of the expression in (4), and obtain the probability distribution of X by using result (3). This is illustrated for *k*= 2, …, 5.

Sum of 2 fair dice (*k* = 2)

$$M_X(t) = \frac{1}{36}\left(e^{2t} + 2e^{3t} + 3e^{4t} + 4e^{5t} + 5e^{6t} + 6e^{7t} + 5e^{8t} + 4e^{9t} + 3e^{10t} + 2e^{11t} + e^{12t}\right) \quad (3.310)$$

which, from result (3.308), is the *mgf*of the following probability distribution.

Table 1: Probability distribution of the sum of 2 fair dice

x	2	3	4	5	6	7	8	9	10	11	12
f(x)	$\frac{1}{36}$	$\frac{2}{36}$	$\frac{3}{36}$	$\frac{4}{36}$	$\frac{5}{36}$	$\frac{6}{36}$	$\frac{5}{36}$	$\frac{4}{36}$	$\frac{3}{36}$	$\frac{2}{36}$	$\frac{1}{36}$

The mgf $M_X(t)$ of the sum S for *k* = 3, 4, 5 is similarly obtained from equation (4):

Sum of 3 fair dice (*k* = 3)

$$M_X(t) = \frac{1}{216}\begin{pmatrix} e^{3t} + 3e^{4t} + 6e^{5t} + 10e^{6t} + 15e^{7t} + 21e^{8t} + 25e^{9t} + 27e^{10t} \\ +27e^{11t} + 25e^{12t} + 21e^{13t} + 15e^{14t} + 10e^{15t} + 6e^{16t} + 3e^{17t} + e^{18t} \end{pmatrix} \quad (3.311)$$

Sum of 4 fair dice (*k* = 4)

$$M_X(t) = \frac{1}{1296}\begin{pmatrix} e^{4t} + 4e^{5t} + 10e^{6t} + 20e^{7t} + 35e^{8t} + 56e^{9t} + 80e^{10t} + 104e^{11t} + \\ 125e^{12t} + 140e^{13t} + 146e^{14t} + 140e^{15t} + 125e^{16t} + 104e^{17t} + \\ 80e^{18t} + 56e^{19t} + 35e^{20t} + 20e^{21t} + 10e^{22t} + 4e^{23t} + e^{24t} \end{pmatrix} \quad (3.312)$$

Sum of 5 fair dice ($k = 5$)

$$M_X(t)=\frac{1}{7776}\begin{pmatrix} e^{5t}+5e^{6t}+15e^{7t}+35e^{8t}+70e^{9t}+126e^{10t}+205e^{11t}+ \\ 305e^{12t}+420e^{13t}+540e^{14t}+651e^{15t}+735e^{16t}+780e^{17t}+ \\ 780e^{18t}+735e^{19t}+651e^{20t}+540e^{21t}+420e^{22t}+305e^{23t}+ \\ 205e^{24t}+126e^{25t}+70e^{26t}+35e^{27t}+15e^{28t}+5e^{29t}+e^{30t} \end{pmatrix} \quad (3.313)$$

The probability distributions of the sum of k dice for k = 3, 4, and 5 are shown in the following table.

Table 2: Probability distribution of the sum of k dice, $k = 3, 4, 5$.

k=3		k=4		k=5	
Sum x	$f(x)$	Sum x	$f(x)$	Sum x	$f(x)$
3	1/216	4	1/1296	5	1/7776
4	3/216	5	4/1296	6	5/7776
5	6/216	6	10/1296	7	15/7776
6	10/216	7	20/1296	8	35/7776
7	15/216	8	35/1296	9	70/7776
8	21/216	9	56/1296	10	126/7776
9	25/216	10	80/1296	11	205/7776
10	27/216	11	104/1296	12	305/7776
11	27/216	12	125/1296	13	420/7776
12	25/216	13	140/1296	14	540/7776
13	21/216	14	146/1296	15	651/7776
14	15/216	15	140/1296	16	735/7776
15	10/216	16	125/1296	17	780/7776
16	6/216	17	104/1296	18	780/7776
17	3/216	18	80/1296	19	735/7776
18	1/216	19	56/1296	20	651/7776
		20	35/1296	21	540/7776
		21	20/1296	22	420/7776
		22	10/1296	23	205/7776

		23	4/1296	24	205/7776
		24	1/1296	25	126/7776
				26	70/7776
				27	35/7776
				28	15/7776
				29	5/7776
				30	1/7776

Example 3.20

This is another example of the application of the Laplace transform from statistics. In this example, the prior density of Mean Time Before Failure (MTBF) is estimated via Laplace transform inversion when the failure times of items put on a life test have a conditional exponential distribution[58]

$$f(x\,|\,\lambda) = \lambda e^{-\lambda x},\; x > 0 \tag{3.314}$$

and the hazard rate λis a random variable with a prior pdf g(λ)[59]; in classical statistics, the hazard rate λis assumed to be an unknown constant.

Under the above assumptions, the marginal cumulative distribution function of time to failure X is given by

$$1 - F(x) = \bar{F}(x) = \int_0^\infty e^{-\lambda x} g(\lambda)\, d\lambda \tag{3.315}$$

which is the Laplace transform of the prior pdf g(λ). If the prior pdf g(λ) is continuous and is of exponential order as λ→∞, i.e. for some nonnegative constants M, c and T:

$$|g(\lambda)| \le Me^{c\lambda},\; \lambda \ge T.$$

The above equation is a Fredholm integral equation of the first kind and is known to be ill-conditioned., and the Laplace transform 1-*F*(*x*) to be inverted has to be estimated from the data. As such, there is no hope of analytically inverting the Laplace transform and we find a smooth approximation.

In order to get a finite range of integration, we use the substitution $u = e^{-\lambda}$ and obtain

$$\bar{F}(x) = \int_0^1 u^{x-1} g(-\ln u)\, du \tag{3.316}$$

---

58 Ashok K. Singh and Anita Singh (1986). On Estimation of the Prior Density in the Exponential Case by Numerical Inversion of the Laplace Transform. Integral Methods in Science and Engineering, Editors: Fred R. Payne, Constantin C. Corduneanu, A. Haji-Sheikh, Tseng Huang, Hemisphere Publishing Inc., Washington.

59 Samir K. Bhattacharya (1967). Bayesian Approach to Reliability and Life Testing. Journal of the American Statistical Association, Vol. 62, pp. 48-62.

The complimentary cdf is estimated from sample by

$$\hat{\bar{F}}(y)=1-\hat{F}(x)=\frac{\text{number of observations } x_i>y}{n}. \tag{3.317}$$

The integral in (3.315) can be estimated by (i) a suitable numerical quadrature or (ii) by assuming an approximating form for the solution.

(i) Numerical Quadrature

The integral is approximated by

$$1-\hat{F}(x)=\sum_{i=1}^{m} w_i u_i^{x-1} g(-\ln u_i) \tag{3.318}$$

where the nodes u_i are the zeroes of the shifted Legendre polynomials, and w_i are the weights[60]. It is assumed that $g(-\ln u)$ can be reasonably approximated by a polynomial in u for $0\le u\le 1$, or equivalently, $g(u)$ can be approximated by a linear combination of powers of e^{-u} on the interval $[0,\infty)$. A sufficient condition for this is continuity of $g(u)$ on $[0,\infty)$ along with the condition $g(u)\xrightarrow{u\to\infty}0$.

Letting y assume $p > m$ distinct values $y_1, y_2, \ldots, y_p$, the following over-determined linear system of equations is obtained:

$$\sum_{i=1}^{m} b_i u_i^{y_j-1}=h_i \;;\; j=1,2,\ldots,p \tag{3.319}$$

where

$$h_i = 1-\hat{F}(y_i)$$
$$b_i = w_i g(-\ln u_i).$$

We treat the above over-determined system as a linear model and find the vector $\mathbf{b}=(b_1,b_2,\ldots,b_m)$ such that $\|\mathbf{h}-V\mathbf{b}\|$ is minimum, where $\|\cdot\|$ is a vector norm and V is a $p\times m$ matrix given by

---

60 R. Bellman, R.E. Kalaba, and J.A. Lockett (1966). Numerical Inversion of the Laplace Transform, pp. 1-48. American Elsevier Publishing Company, Inc., New York.

$$V_{ji} = u_i^{x_j - 1}, j = 1, 2, \ldots, p; i = 1, 2, \ldots, m. \tag{3.320}$$

If ℓ_2 −norm is used, then **b** is given by[61]

$$\mathbf{b} = (V^T V)^{-1} V^T \mathbf{h} \tag{3.321}$$

where V^T is the transpose of the matrix V.

Since the matrix $V^T V$ is ill-conditioned, we have used the method of ridge regression[51] for which

$$\hat{\mathbf{b}} = (V^T V + kI)^{-1} V^T \mathbf{h} \tag{3.322}$$

where k is a positive constant that is selected by the method of ridge trace.

If ℓ_∞ −norm is used, then **b** can be estimated by the simplex method of linear programming. We will refer to this estimate as min-max estimate.

If ℓ_1 −norm is used, then **b** can be estimated by the method of Least Absolute Deviation (LAD) regression.

The estimated prior at the selected nodes u_i is computed from

$$\hat{g}(-\ln u_i) = \frac{b_i}{w_i}. \tag{3.323}$$

II. Polynomial Approximation

If the prior pdf $g(-\ln u)$ can be approximated as

$$g(-\ln u_i) = \sum_{j=1}^{m} a_j u^j \tag{3.324}$$

then the general linear model becomes

$$\sum_{j=1}^{m} \frac{a_j}{x_i + j} = 1 - \hat{F}(x_i),\ i = 1, 2, \ldots, p. \tag{3.325}$$

---

61 R.F. Gunst and R.L. Mason (1980). Regression Analysis and its Applications, pp. 340-348, Marcel Dekker, Inc., New York.

The regression coefficients are again estimated by the methods of ridge regression, min-max regression, and LAD regression. The prior pdf in this case is estimated by

$$\hat{g}(\lambda)=\sum_{j=1}^{m}\hat{a}_j e^{-j\lambda}. \tag{3.326}$$

We next present results from a few simulated examples.

Simulation Examples

The simulation experiment used to generate examples is described below:

1. A random hazard rate $\lambda_i\left(i=1, 2, ..., n\right)$ is generated from the natural conjugate gamma prior

$$g(\lambda)=\frac{c^d}{\Gamma(d)}\lambda^{d-1}e^{-c\lambda},\lambda>0;\ c>0,\ d>0. \tag{3.327}$$

2. For each $i=1, 2, ..., n,$ a random number x_i is generated from an exponential distribution with hazard rate λ_i.

3. A set of nodes $y_1,y_2,...,y_p$ in the y-space is selected, and the empirical cdf $\hat{F}(y_i)$ is computed for each $j=1, 2, .., p$.

For the method of Numerical Quadrature:

4a. For a fixed *m*, the design matrices for the linear models are calculated.

5a. The regression coefficients b_i and a_i are estimated for the three estimation methods described above.

We have used $m=5,n=200,p=10$.

Example 1: In this example, the true prior is taken to be the natural conjugate gamma prior with $c=d=1$. The estimated priors are shown in the Figure 3.1. It can be seen that the Ridge estimates come closest to the true prior.

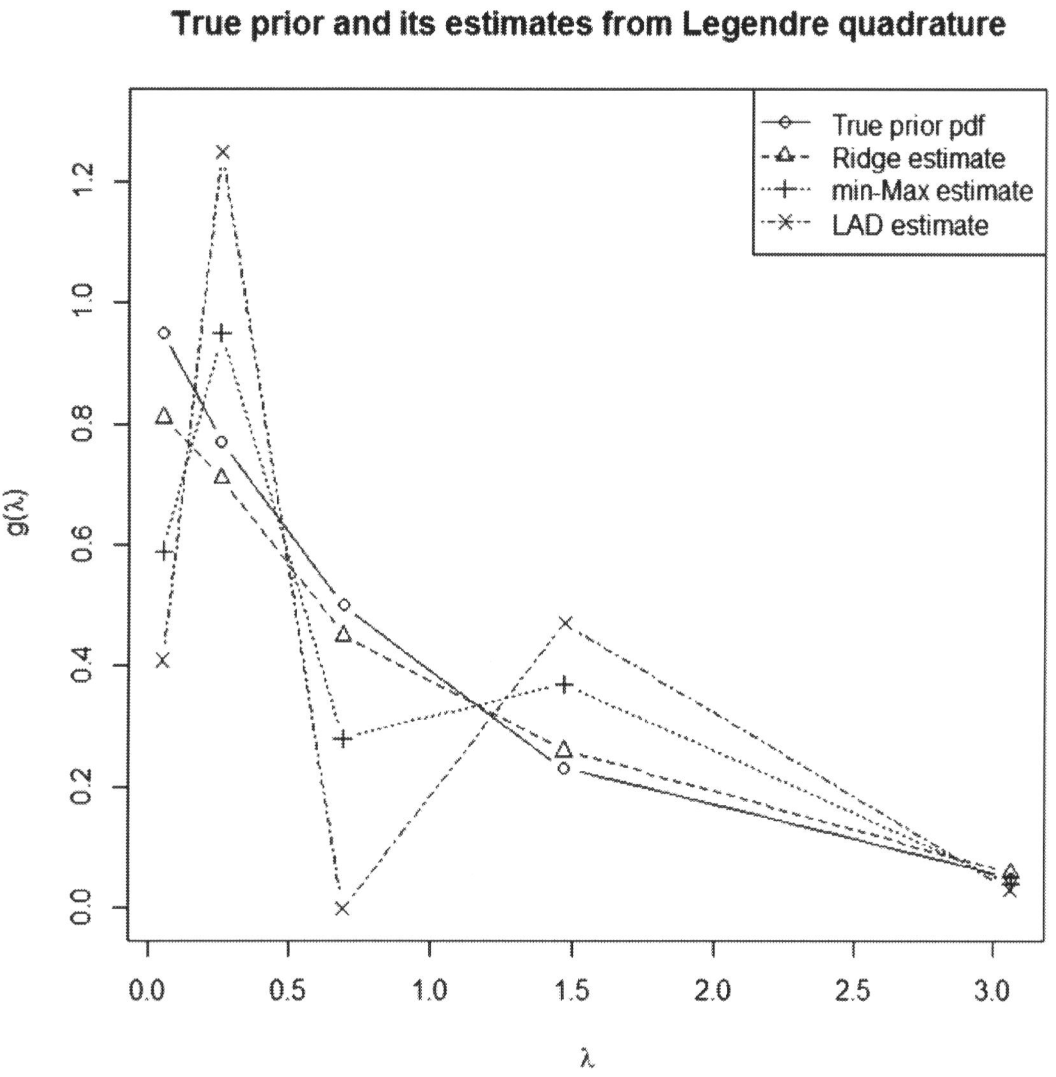

Figure 3.1: True prior and prior estimated from Legendre quadrature for Example 1

Using the method of Polynomial Approximation:

4b. The estimated $g(\lambda)$ turned out to be

$$\hat{g}_{m-M}(\lambda) = \hat{g}_{LAD}(\lambda) = g(\lambda) = e^{-\lambda}, \lambda > 0$$
$$\hat{g}_{R}(\lambda) = 0.72e^{-\lambda} + 0.25e^{-2\lambda} + 0.11e^{-3\lambda} + 0.05e^{-4\lambda} + 0.02e^{-5\lambda}, \lambda > 0$$

It can be seen from the plot of the above four functions that the min-max and LAD estimates have no error, and also that the ridge estimate is also quite close to the true $g(\lambda)$.

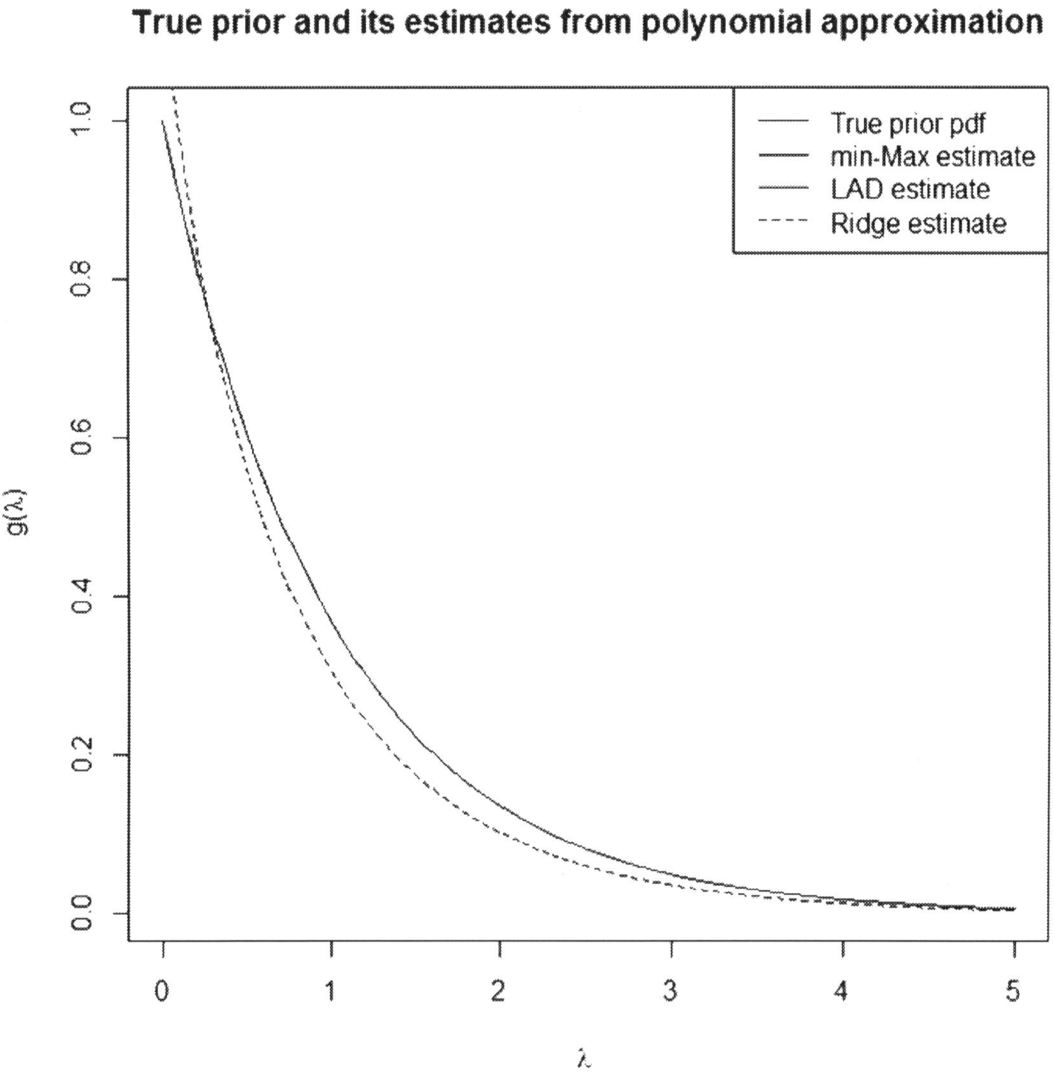

Figure 3.2: True and prior estimated from polynomial approximation for Example 1

Example 2: In this example, the true prior is taken to be the natural conjugate gamma prior with c = 2, d = 1. The results obtained from the numerical quadrature approximation are shown in Figure 3.3.

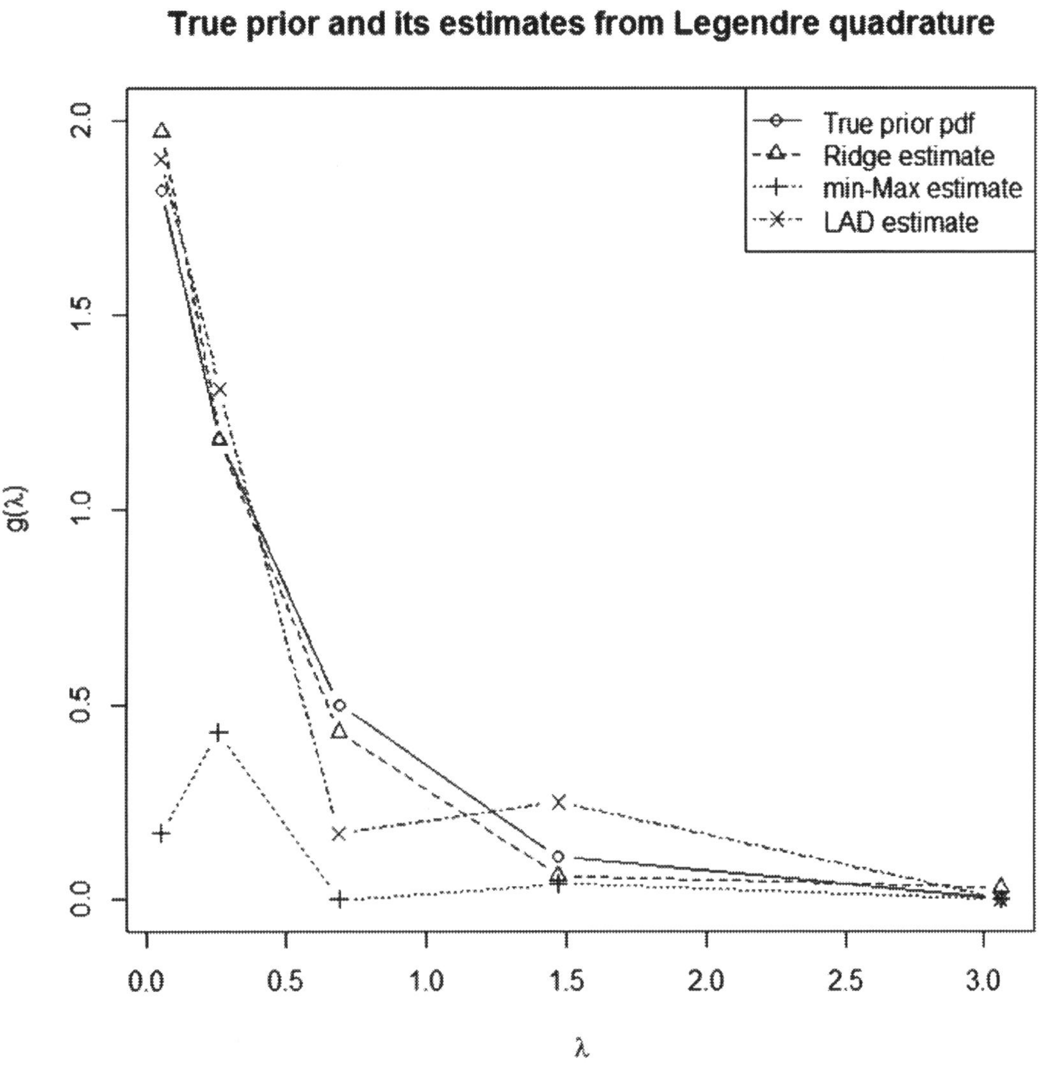

Figure 3.3: True prior and prior estimated from Legendre quadrature for Example 2

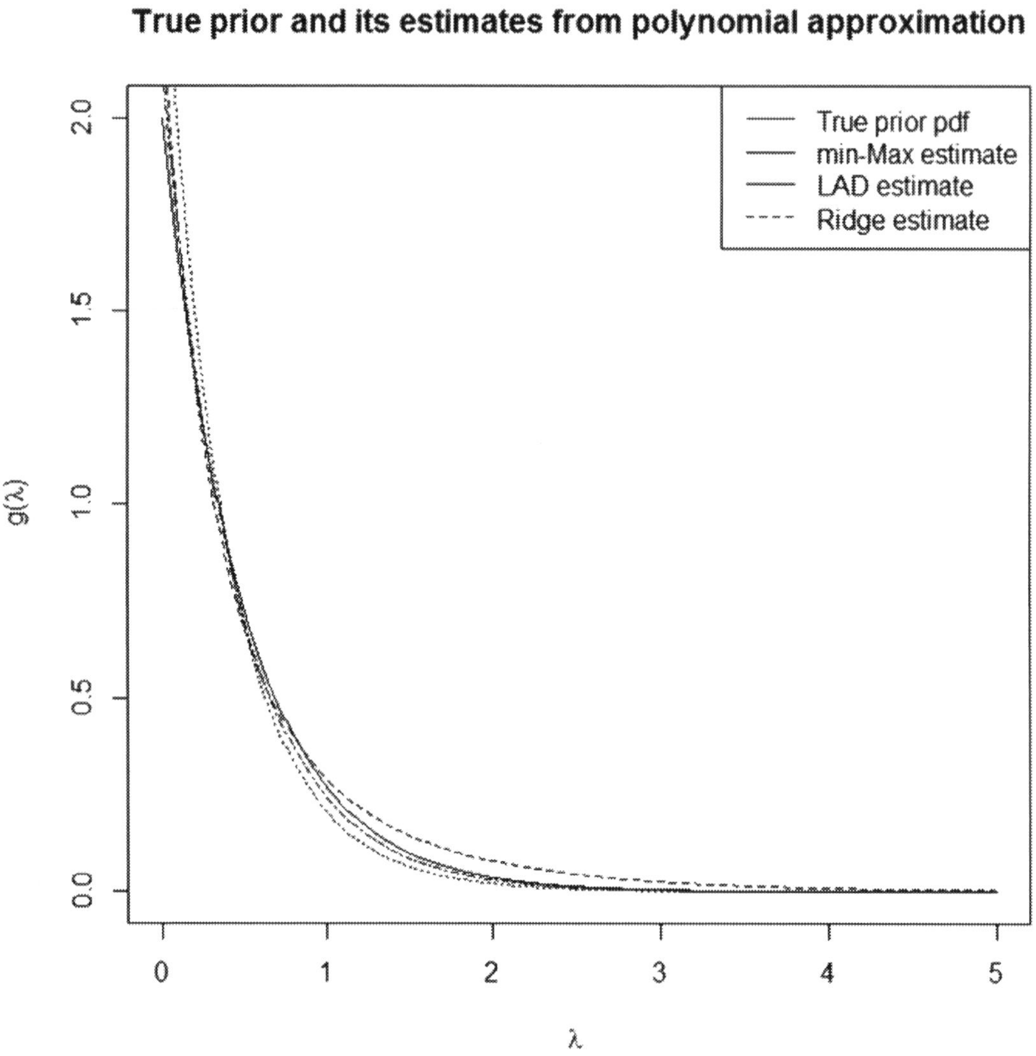

Figure 3.4: True and prior estimated from polynomial approximation for Example 2

It appears from Figure 3.4 that the method of polynomial approximation produces reasonable estimates of the prior distribution $g(\lambda)$.

4. NUMERICAL INVERSION OF THE LAPLACE TRANSFORM

The Laplace transform given by equation (2.1) is a special case of the Fredholm integral equation of the first kind:

$$\mathcal{L}\{f\} = F(s) = \int_a^b K(s,t)f(t)dt \qquad (4.1)$$

where $K(s,t)$ is the kernel.

It is a well known fact that the problem of solving the above Fredholm integral equation is ill-conditioned[62]. There are numerous situations in which solution to a problem can be found in the Laplace domain, but the analytical inversion of the Laplace transform may be too complicated. A large number of numerical methods exist for inversion of the Laplace transform[63],[64]; as to be expected, no 'best' method exists.

4.1: Gaver-Stehfest Method

The Gaver-Stehfest method[65] is a popular numerical inversion method, which gives good accuracy on a wide range of functions. The numerical inversion formula of Gaver-Stehfest is given below:

$$f(t) = \frac{\ln(2)}{t} \sum_{n=1}^{N} c_n F\left(\frac{n\ln(2)}{t}\right) \qquad (4.2)$$

62 J.G. McWhirter and E.R. Pike (1978). Journal of Physics A: Mathematics General, Vol. 11, pp. 1729 - 1745.

63 Brian Davies and Brian Martin (1979). Numerical Inversion of the Laplace Transform : a Survey and Comparison of Methods. Journal of Computational Physics 33, pp. l-32.

64 Alan M. Cohen (2007). Numerical Methods for Laplace Transform Inversion. Springer Science & Business Media.

65 Alexander H-D. Cheng, PastonSidauruk, YounaneAbousleiman (1994). Mathmatica Journal, pp. 76-82.

where the coefficients c_n are computed from

$$c_n = (-1)^{n+N/2} \sum_{k=L(n)}^{U(n)} \frac{k^{N/2}(2k)!}{(N/2-k)!k!(k-1)!(n-k)!(2k-n)!}$$

with $L(n) = [(n+1)/2]$, $U(n) = \min(n, N/2)$, $[a] =$ largest integer $\le a$.

We include a code in the programming language R[66] for the Gaver-Stehfest algorithm.

The Gaver-Stehfest algorithm, commonly used to solve problems in Hydrology and Petroleum Engineering, has also been used to numerically invert the moment generating functions to compute the probability distributions of sums of independent random variables[67].

We next provide the results of the accuracy of Laplace transform inversion method of Gaver-Stehfest on the test functions (Laplace transform pairs) shown in Table 4.1.

Table 4.1: Laplace transform pairs used as test functions

Laplace Transform	Exact Inverse
$F_1(p) = 1/\sqrt{p}$	$f_1(x) = 1/\sqrt{\pi x}$
$F_2(p) = \log(p)/p$	$f_2(x) = -0.57722 - \ln(x)$
$F_3(p) = 1/(p+1)$	$f_3(x) = e^{-x}$
$F_4(p) = e^{-\frac{1}{2x}}\sqrt{\pi/2p^3}$	$f_4(x) = \sin(\sqrt{2x})$
$F_5(p) = 1/p$	$f_5(x) = 1$
$F_6(x) = \sqrt{\pi}/2x^{3/2}$	$f_6(x) = \sqrt{x}$
$F_7(p) = \frac{e^{-1/p}}{\sqrt{x}}$	$f_7(x) = \frac{\cos(2\sqrt{x})}{\sqrt{\pi x}}$

66 R Core Team (2014). R: A language and environment for statistical computing. R Foundation for Statistical Computing, Vienna, Austria, URL http://www.R-project.org/.

67 Andy Tsang (1997). On Numerical Inversion of the Moment Generating Function. M.S. Thesis, Department of Mathematical Sciences, University of Nevada, Las Vegas.

$F_8(p) = \tan^{-1}\left(\frac{1}{x}\right)$	$f_8(x) = \frac{\sin(x)}{x}$

Figures 4.1 and 4.2 show that exact inverse and the numerically computed inverse using the Gaver-Stehfest method. For all of the above test functions, the Gaver-Stehfest is seen to provide accurate results over the range of x values considered.

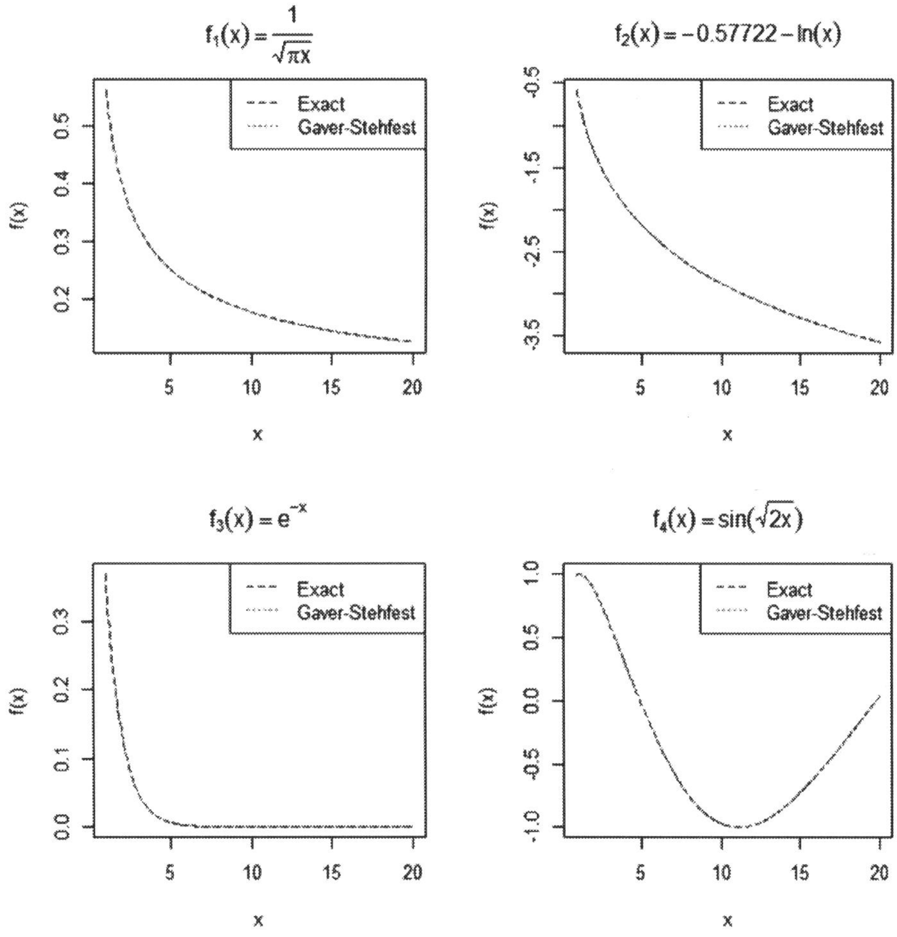

Figure 4.1: Exact and numerically compute inverse functions for test functions 1 - 4

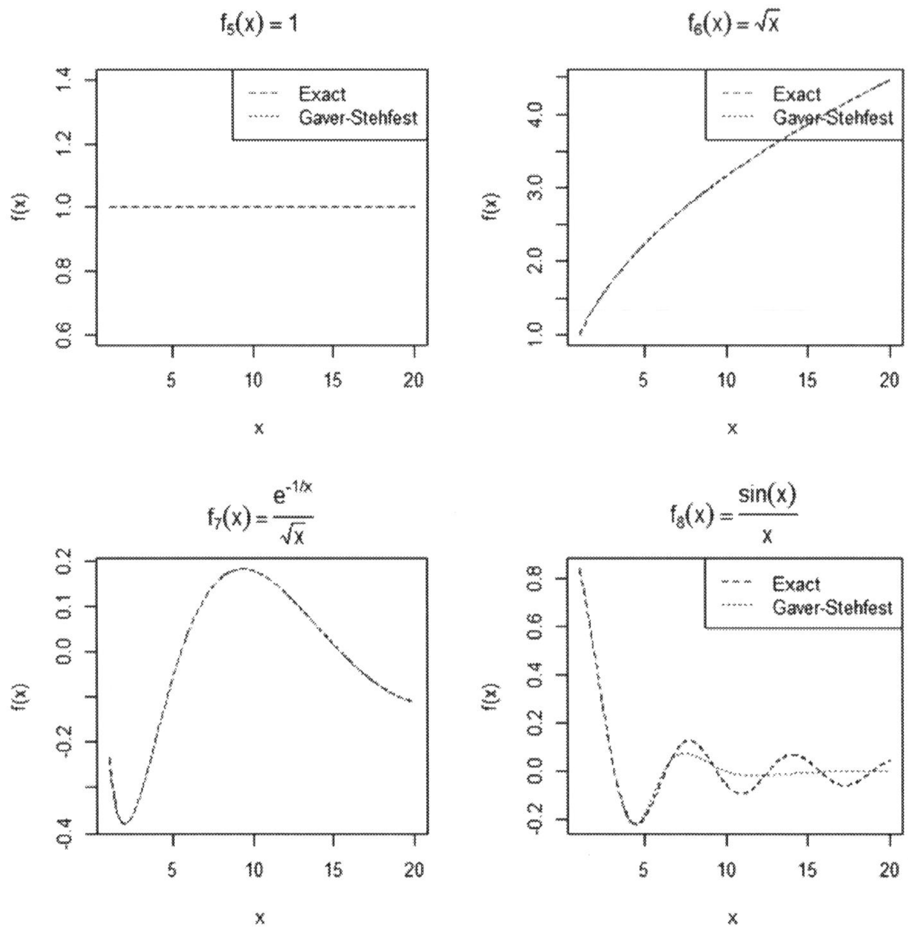

Figure 4.2: Exact and numerically compute inverse functions for test functions 5 - 8

For each of the eight test functions, we have computed three measures of numerical accuracy:

$$E1 = \sqrt{\sum_{k=1}^{40} \frac{\left(f(k) - f_{GS}(k)\right)^2}{40}}$$

$$E2 = \sqrt{\frac{\sum_{k=1}^{40} \left(f(k) - f_{GS}(k)\right)^2 e^{-k}}{\sum_{k=1}^{40} e^{-k}}}$$

$$E3 = \max\left(\left|f(k) - f_{GS}(k)\right|\right)$$

where $f(k)$ is the exact Laplace inverse and $f_{GS}(k)$ the approximate Laplace inverse obtained from the Gaver-Stehfest algorithm. The measure E_1 gives a fair indication of accuracy for large t values, and E_2 for small t values[68]. In all the accuracy measure calculations, t = .01, 0.02, ..., 5.00. Table 4.2 shows the three errors computed for the eight test Laplace transform pairs of Table 4.1.

Table 4.2: Accuracy measures for numerical inversion of the eight test functions of Table 4.1

Test function	E1	E2	E3
F1(p)	0.000000	0.000000	0.000001
F2(p)	0.000004	0.000000	0.000005
F3(p)	0.000010	0.000000	0.000019
F4(p)	0.000266	0.000000	0.000573
F5(p)	0.000000	0.000000	0.000000
F6(p)	0.000000	0.000000	0.000000
F7(p)	0.000509	0.000000	0.001123
F8(p)	0.040441	0.000000	0.078535

It can be seen from Table 4.2 that, except for the last test function, the method of Gaver-Stehfest gives good accuracy.

In Section 4.2, we provide a computer code in the programming language R[69].

4.2: An R code for the Method of Gaver-Stehfest

R is a free software for mathematical and statistical computing and data visualization. There are numerous free manuals and online courses available for R, but for the sake of completeness, we have included a section on basics of computing and visualization in R. In this section, we provide an R code for numerically inverting a given Laplace transform by the method of Gaver-Stehfest.

68 Brian Davies and Brian Martin (1979). Numerical Inversion of the Laplace Transform : a Survey and Comparison of Methods. Journal of Computational Physics 33, pp. l-32.

69 R Core Team (2014). R: A language and environment for statistical computing. R Foundation for Statistical Computing, Vienna, Austria. URL http://www.R-project.org/.

```
##Begin R Code for Laplace Transform Inversion by Stehfest Algorithm ############
# Note: Any line starting with # is a comment line in R.
# ------------------------------------------------------------------------------------------------------------------------
# Since the inversion of the Laplace transform is an ill-conditioned problem, we compute
# the coefficients of equation (3.328) in double precision which are saved in a .csv file.
# This file is used in Step 2.
# ------------------------------------------------------------------------------------------------------------------------
# Step 1: calculate all of the coefficients of (3.328) using double precision arithmetic
# and N = 16 (N must be an even integer)
N <- 16
Nh <- N/2
c <- vector()
b <- vector()
L <- vector()
U <- vector()

for (n in 1:N)
{
b[n] <- as.double((-1)**(n+Nh))
L[n] <- floor((n+1)/2)
U[n] <- min(n, Nh)
c[n] <- as.double(0.0)
for (k in L[n]:U[n])
{num <- as.double((k**Nh)*factorial(2*k))
den <- as.double(factorial(Nh-k)*factorial(k)*factorial(k-1)*factorial(n-k)*factorial(2*k-n))
c[n] <- c[n] + b[n]*num/den
}
}
write.csv(c,"cnDouble.csv") # saves the coefficients in a .csv file.

#Step 2: funcion Lap.inverse ################################################
Lap.inverse <- function(F,t)
{
Y <- read.csv("cnDouble.csv",header=TRUE) # read the coefficients computed in Step 1
c <- Y$x
```

```
terms <- vector()
N <- 16
to <- log(2)/t
for (n in 1:N) terms[n] <- to*c[n]*F(n*to)
result <- sum(terms)
return(result)
}
# Step 3 Invert given Laplace transform F ########################################
# Step 3(a): type the Laplace Transforms to be inverted
# F(x) = the test Laplace Transforms to be inverted
# Laplace transform F(x) = 1/sqrt(x), exact inverse f1(t) = 1/sqrt(pi*t)

F <- function(x)
{
result <- 1/sqrt(x)
return(result)
}

#Step 3(b): Use the function Lap.inverse to compute the approximate inverse
t <- seq(0.01,5, by = 0.01)
for (i in 1:length(t)) fapprox[i] <- Lap.inverse(F,t[i])
print(fapprox) # this will show the numerical inverse in R-console window.
# fapprox can be saved in a .csv file as shown in Step 1 above.

# The vector fapprox has the numerical inverse o at t-values.01, 0.02, ..., 5.00
# End of the R Code for Laplace Transform Inversion by Stehfest Algorithm #######
```

To run this code in R, follow the steps given below:

1) Start an R session (see Section 4.3).

2) In R, click on File (top left corner), then on Open Script, which will open the R-editor window.

3) Copy the code shown above (from Begin to End) and paste into the editor window. Type the Laplace transform that you need to invert in Step 3(a), and then you can run the entire code all at once by clicking on Edit and then on 'Run all'.

4.3: Introduction to R

| 4.0.1 DOWNLOAD R

1) Google R, then click on the link 'R: The R Project for Statistical Computing', which will take you to 'http://www.r-project.org/'; you will see a window similar to the one shown below.

The R Project for Statistical Computing

[Home]

Download

CRAN

R Project

About R
Contributors
What's New?
Mailing Lists
Bug Tracking
Conferences
Search

R Foundation

Foundation
Board
Members

Getting Started

R is a free software environment for statistical computing and graphics. It compiles and runs on a wide variety of UNIX platforms, Windows and MacOS. To **download R**, please choose your preferred CRAN mirror.

If you have questions about R like how to download and install the software, or what the license terms are, please read our answers to frequently asked questions before you send an email.

News

- R 3.2.1 (World-Famous Astronaut) prerelease versions will appear starting June 8. Final release is scheduled for 2015-06-18.

- **R version 3.2.0** (Full of Ingredients) has been released on 2015-04-16.

- **R version 3.1.3** (Smooth Sidewalk) has been released on 2015-03-09.

- **The R Journal Volume 6/2** is available.

- **useR! 2015**, will take place at the University of Aalborg, Denmark, June 30 - July 3, 2015.

- **useR! 2014**, took place at the University of California, Los Angeles, USA June 30 - July 3, 2014.

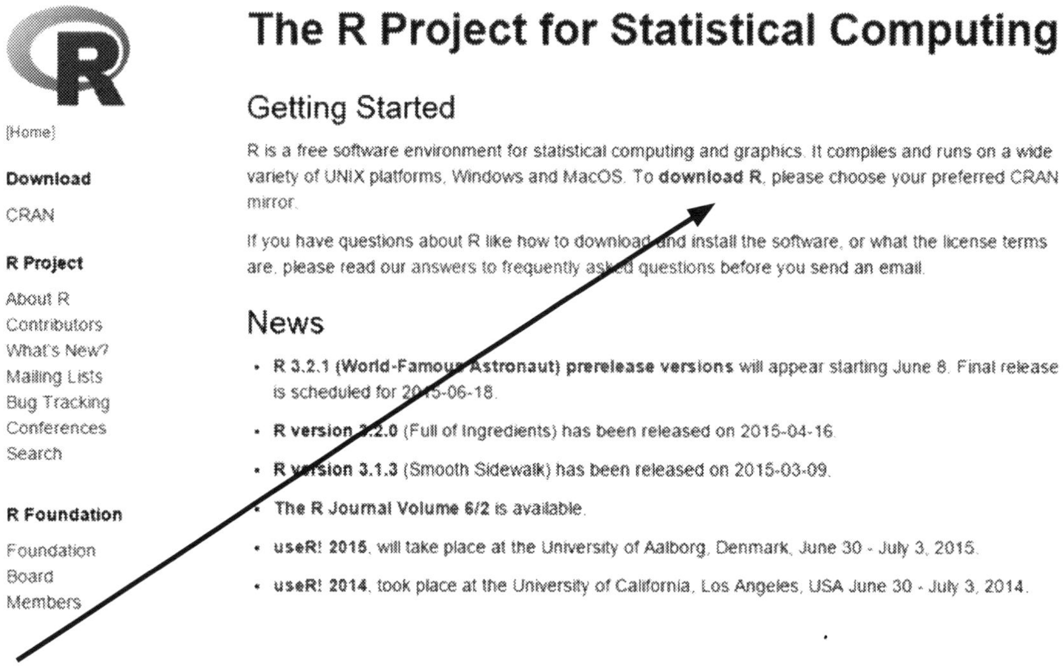

2) Click on 'download R', which will take you to a list of CRAN Mirror sites; you can click on any of the sites. As an example, click on '0-Cloud' link.

The Comprehensive R Archive Network is available at the following URLs. please choose a location close t

0-Cloud
- http://cran.rstudio.com/

Algeria
- http://cran.usthb.dz/

Argentina
- http://mirror.fcaglp.unlp.edu.ar/CRAN/

Australia
- http://cran.csiro.au/
- http://cran.ms.unimelb.edu.au/

Austria
- http://cran.at.r-project.org/

Belgium
- http://www.freestatistics.org/cran/

UK
- http://www.stats.bris.ac.uk/R/
- http://mirrors.ebi.ac.uk/CRAN/
- http://mirrors-uk2.go-parts.com/cran/
- http://cran.ma.imperial.ac.uk/
- http://mirror.mdx.ac.uk/R/
- http://star-www.st-andrews.ac.uk/cran/

USA
- http://cran.cnr.Berkeley.edu/
- http://cran.stat.ucla.edu/
- http://mirror.las.iastate.edu/CRAN/
- http://ftp.ussg.iu.edu/CRAN/
- http://rweb.quant.ku.edu/cran/
- http://watson.nci.nih.gov/cran_mirror/
- http://cran.mtu.edu/
- http://www.go-parts.com/mirrors-usa/cran/
- http://cran.wustl.edu/
- http://cran.case.edu/
- http://iis.stat.wright.edu/CRAN/
- http://ftp.osuosl.org/pub/cran/
- http://lib.stat.cmu.edu/R/CRAN/
- http://cran.mirrors.hoobly.com/
- http://mirrors.nics.utk.edu/cran/

3) Clicking on the link to a site will show you the following:

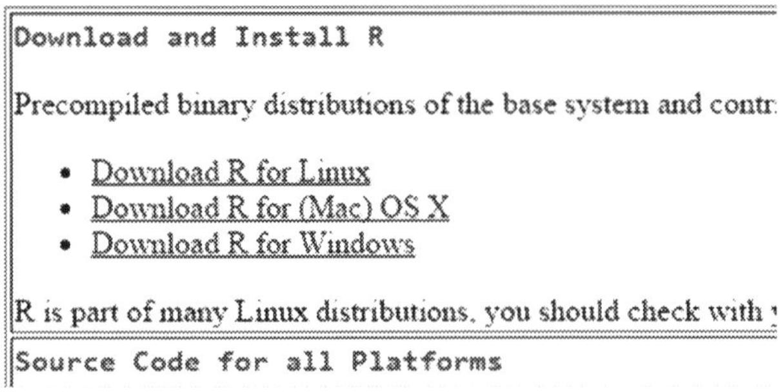
Download and Install R
Precompiled binary distributions of the base system and contr
Download R for Linux
Download R for (Mac) OS X
Download R for Windows
R is part of many Linux distributions, you should check with
Source Code for all Platforms

Click on the platform you use to download R for your computer; click on 'install R for the first time' which will install the base R-package.

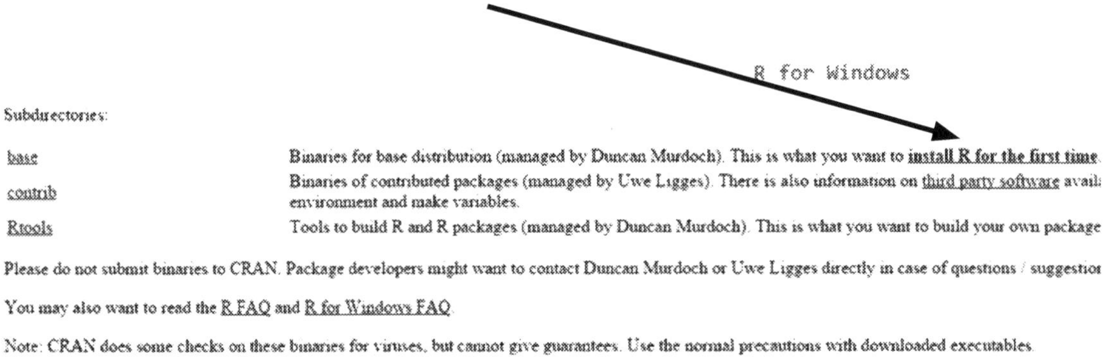
R for Windows
Subdirectories:
base
Binaries for base distribution (managed by Duncan Murdoch). This is what you want to install R for the first time.
contrib
Binaries of contributed packages (managed by Uwe Ligges). There is also information on third party software avail
environment and make variables.
Rtools
Tools to build R and R packages (managed by Duncan Murdoch). This is what you want to build your own package
Please do not submit binaries to CRAN. Package developers might want to contact Duncan Murdoch or Uwe Ligges directly in case of questions / suggestion
You may also want to read the R FAQ and R for Windows FAQ.
Note: CRAN does some checks on these binaries for viruses, but cannot give guarantees. Use the normal precautions with downloaded executables.

To install R for windows platform, click on 'Download R x.y.z for Windows. Select all defaults.

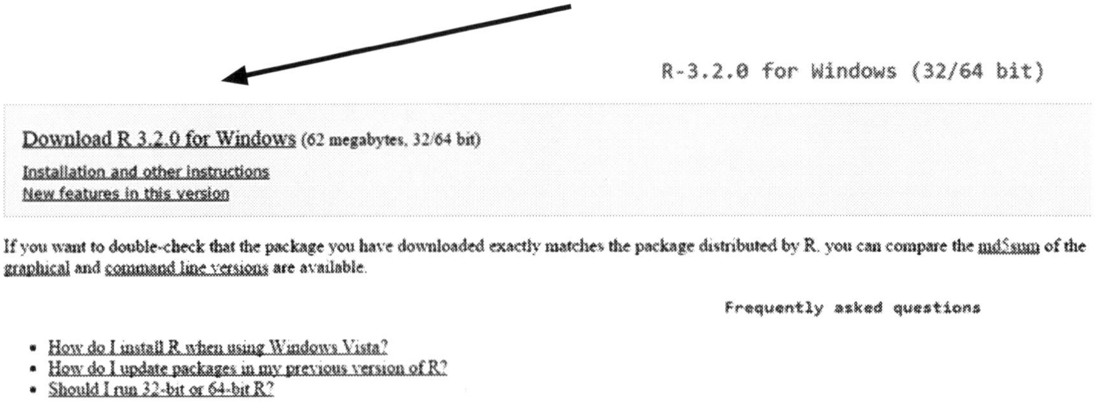
R-3.2.0 for Windows (32/64 bit)
Download R 3.2.0 for Windows (62 megabytes, 32/64 bit)
Installation and other instructions
New features in this version
If you want to double-check that the package you have downloaded exactly matches the package distributed by R, you can compare the md5sum of the
graphical and command line versions are available.
Frequently asked questions
How do I install R when using Windows Vista?
How do I update packages in my previous version of R?
Should I run 32-bit or 64-bit R?

4) The initial download gives you a large number of functions in the base package of R; you can go to 'https://stat.ethz.ch/R-manual/R-devel/library/base/html/00Index.html' to see a

complete listing. For advanced statistical and mathematical computing, you may need to install a package (available from the internet for free). Most of the times, you can google the topic to find the name of the package you need.

5) Once the download is complete and R is icon is stalled on your computer, you will see the following icon on your computer.

6) To start an R session, double-click this icon. Click on File then on 'New script' to open an editor window.

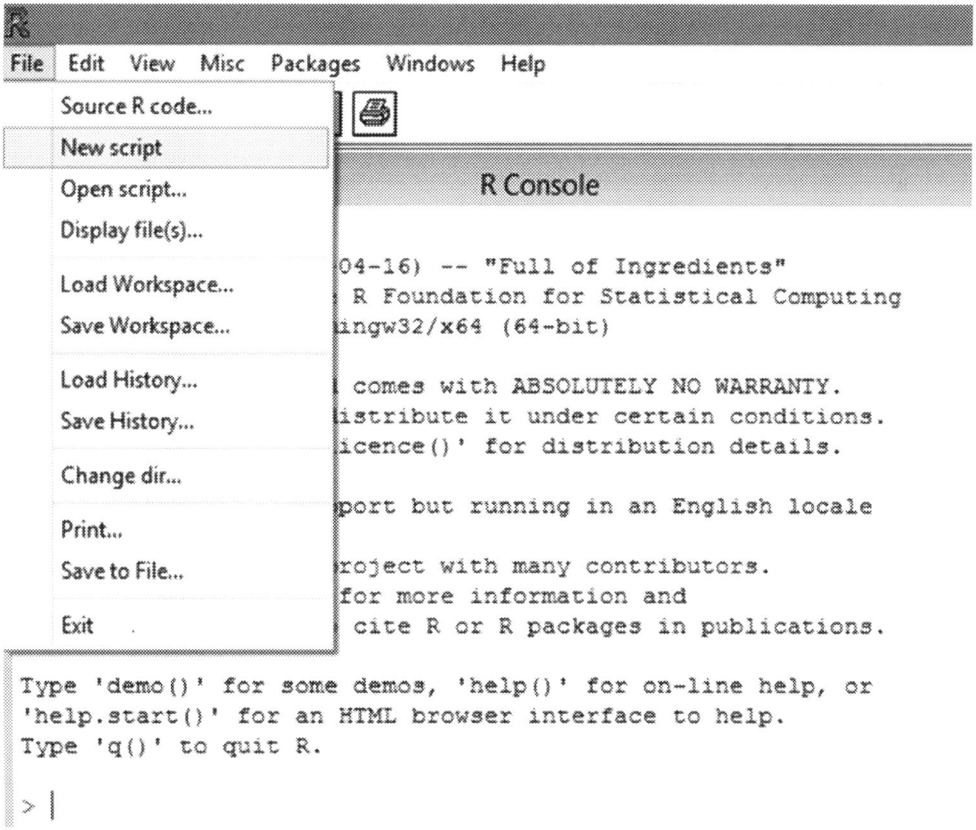

You should now have an 'R Console' window (all results and error messages are shown here) and an untitled R 'Editor window'.

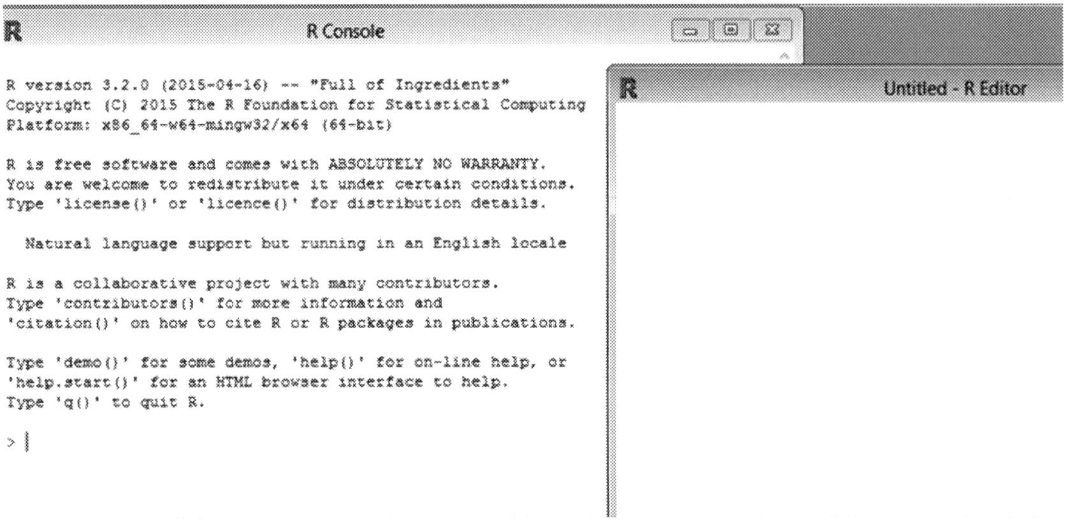

7) Type the following lines of code in the editor window:

Program 'Hi there.R'

temp <- "Hi there"

print(temp)

8) To run your first code, highlight all lines, right click then on 'Run line or selection'; if there are no typographical errors, this will show the following.

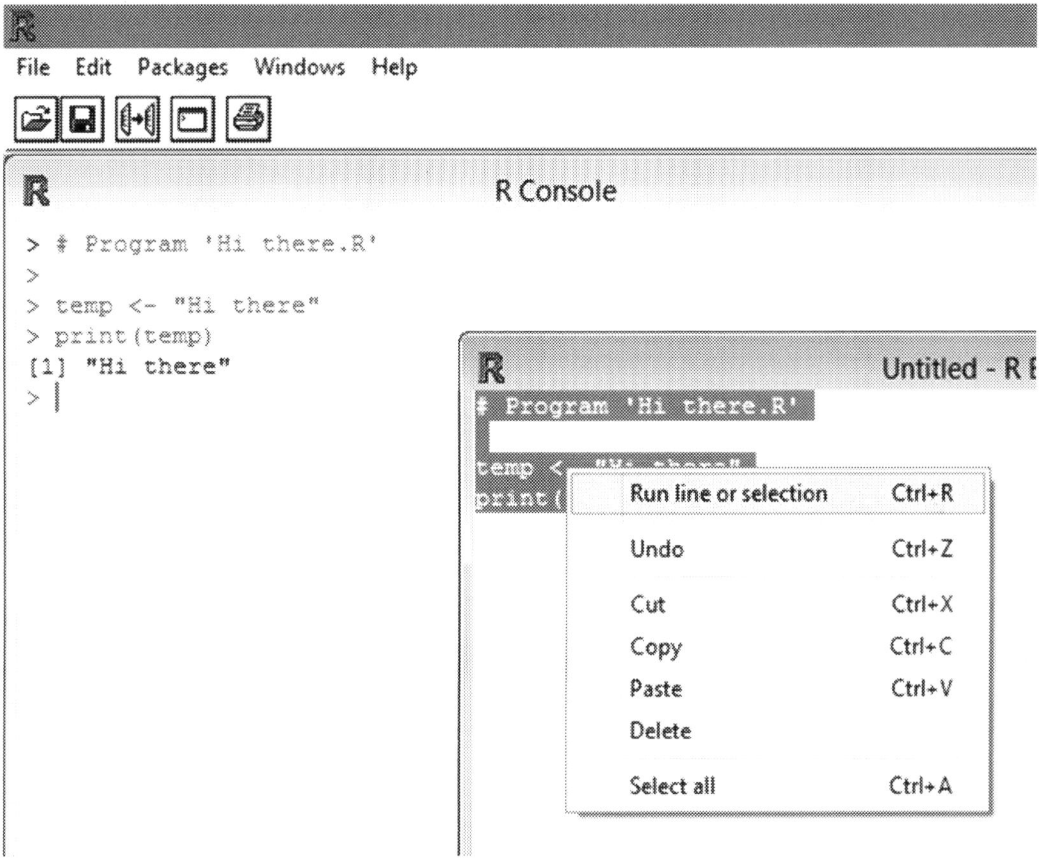

You can save the code by clicking on File and then 'Save as'.

| 4.0.2 COMPUTATIONS IN R

In the following, the result shown in the console window has been *italicized.*

1) You can use R as a scientific calculator by typing expressions directly in the R-Console.

Example 1(a): To calculate the area of circle of radius, type the following three lines in console window one-by-one: the last line prints out the area of the unit circle.

```
> r <- 1
> Area <- pi*r**2
```

> *Area*

[1] 3.141593

2) Matrix calculations

Example 2(a) Small matrices can be directly entered in console or in the editor window.

```
A <- matrix(c(4,3,4,2), nrow=2, byrow=T)
A # or print(A) will show A in the console
B <- matrix(c(1, 1, 2, 2, 3, 3), ncol=3, byrow=T)
B
# A is a 2X2 matrix, B is a 2x3 matrix, so AxB is a 2X3 mtrix.
C <- A%*%B # standard matrix multiplication, C is 2x3 matrix
C1
```

If you highlight the above lines in R-Editor window, right-click, go to File, then select 'Run all', you will get the following in the console:

```
> A <- matrix(c(4,3,4,2), nrow=2, byrow=T)
> B <- matrix(c(1, 1, 2, 2, 3, 3), ncol=3, byrow=T)
> # A is a 2X2 matrix, B is a 2x3 matrix, so AxB is a 2X3 mtrix.
> C <- A%*%B # standard matrix multiplication
>
> A <- matrix(c(4,3,4,2), nrow=2, byrow=T)
> A # or print(A) will show A in the console
  [,1] [,2]
[1,] 4 3
[2,] 4 2
> B <- matrix(c(1, 1, 2, 2, 3, 3), ncol=3, byrow=T)
> B
  [,1] [,2] [,3]
[1,] 1 1 2
[2,] 2 3 3
> # A is a 2X2 matrix, B is a 2x3 matrix, so AxB is a 2X3 mtrix.
> C <- A%*%B # standard matrix multiplication, C is 2x3 matrix
> C
  [,1] [,2] [,3]
[1,] 10 13 17
[2,] 8 10 14
```

3) Solve a simple linear equation:

Example 3(a): To solve 10x = 2, type
solve(10,2)
in console (or editor and run the one-line code) to get the following result:

> solve(10,2)
[1] 0.2

Example 3(b): Solve the following linear system.

$3x + y - 6z = -10$
$2x + y - 5z = -8$
$6x - 3y + 3z = 0$

You can directly enter A in the console (as shown in Example 2 above), or type the matrix of the system in an excel file and save it as a .csv file (say A Example 4.csv)..

3	1	-6
2	1	-5
6	-3	3

```
A <- read.csv("A Example 4.csv", header=FALSE) # since the data file has no column names
# reading A gives a data frame in R, convert to a matrix:
A <- as.matrix(A)
b <- read.csv("RHS.csv",header=FALSE)
b <- as.matrix(b) # convert data frame b to a matrix
solve(A,b)
```

Running this code will result in the following solution:

```
   V1
V1 -2
V2 -4
V3 0
```

Example 3(c) You can invert the above matrix A by typing

solve(A) in the console.

> solve(A)

[,1] [,2] [,3]

V1 -11 14 1

V2 -38 48 3

V3 -12 15 1

Verify that the above matrix is A^{-1}:

A%*%solve(A)

> A%%solve(A)*

[,1] [,2] [,3]

[1,] 1.000000e+00 0 0.000000e+00

[2,] 1.421085e-14 1 -1.776357e-15

[3,] 0.000000e+00 0 1.000000e+00

Example 3(d) Solve the non-linear equation in one unknown $(\cos(2x))^3 = 0$.

Here we will need to install a package called rootSolve. To do so, click on Packages then 'on Install packages', select a CRAN site (or just select cloud)

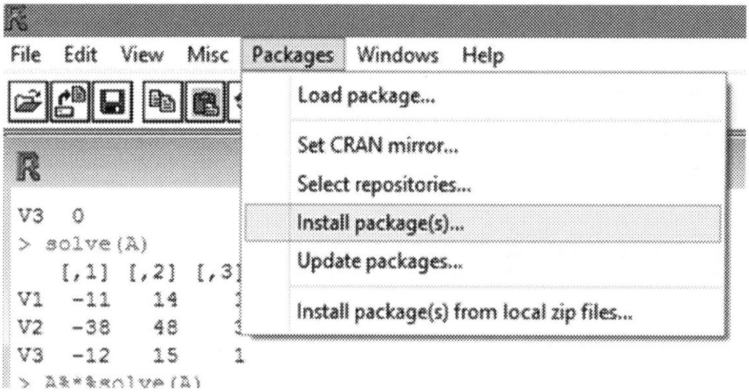

This will open a list of available R-packages. Scroll down and select rootSolve to install this package on your computer. Both of the examples

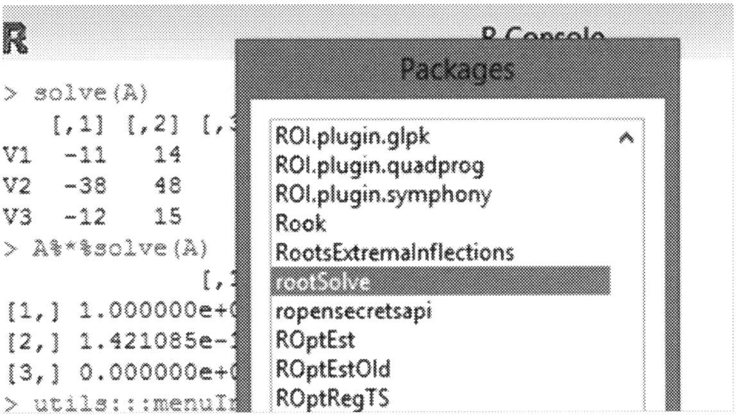

In order to use this package, type the following in R-console (or editor):

library(rootSolve)

This will result in something similar to the following lines in the console window:

> library(rootSolve)
Warning message:
package 'rootSolve' was built under R version 3.1.3

R-code for graphing the above quartic function and then solving the quartic equation of Example 5(a).

```
fun <- function (x) cos(2*x)**3
curve(fun(x),0,8)
All <- uniroot.all(fun, c(0, 8))
```

> All
[1] 0.7853994 2.3561753 3.9270078 5.4977787 7.0685537

points(All, y = rep(0, length(All)), pch = 16, cex = 2) # will plot the 5 roots inside (0,8).

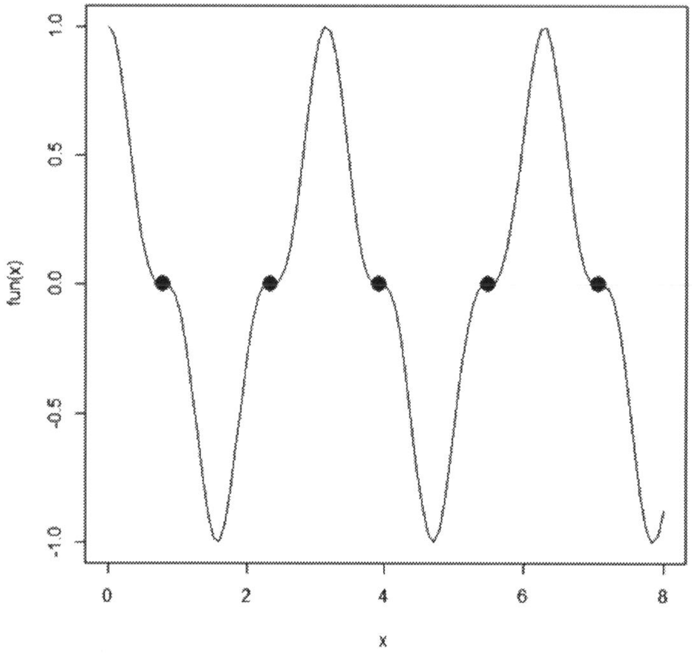

Figure 4.3: Plot of the function of example 3(d) and its 5 roots inside the interval (0,8)

Example 3(e) Solve the non-linear system of equations

$x_1 + x_2 + x_3^2 = 12$

$x_1^2 - x_2 + x_3 = 2$

$2x_1 - x_2^2 + x_3 = 1$

```
F <- function(x) {
F1 <- x[1] + x[2] + x[3]^2 -12
F2 <- x[1]^2 - x[2] + x[3] -2
F3 <- 2*x[1] - x[2]^2 + x[3] -1
result <- c(F1 = F1, F2 = F2, F3 = F3)
return(result)}
# first solution
 (ss <- multiroot(f = F, start = c(1, 1, 1)))
# second solution
 (ss <- multiroot(f = F, start = c(0, 0, 0)))
```

Output is shown on the next page.

Output from R:

```
> # first solution
> (ss <- multiroot(f = F, start = c(1, 1, 1)))
$root
[1] 1 2 3
$f.root (function value at the root)
      F1        F2        F3
3.087877e-10 4.794444e-09 -8.678146e-09
$iter
[1] 6
$estim.precis
[1] 4.593792e-09
>
> # second solution
> (ss <- multiroot(f = F, start = c(0, 0, 0)))
$root
[1] -0.2337207 1.3531901 3.2985649
$f.root
      F1        F2        F3
1.092413e-08 1.920978e-07 -4.850423e-08
$iter
[1] 10
$estim.precis
[1] 8.384205e-08
```

Both of the examples 5(a) and 5(b) are from the vignette for rootSolve[70].

---

70 http://cran.r-project.org/web/packages/rootSolve/vignettes/rootSolve.pdf

4.4: Introduction to RStudio

RStudio is an Integrated Development Environment (IDE) which works with the R application installed on your computer to provide additional functionality.

1) After downloading R, download RStudio from http://www.rstudio.com/

2) After RStudio has been downloaded and installed, click on the R-studio icon:

This will open the RStudio GUI with four windows: Editor window, Workspace window, Console window, and Graphics window.

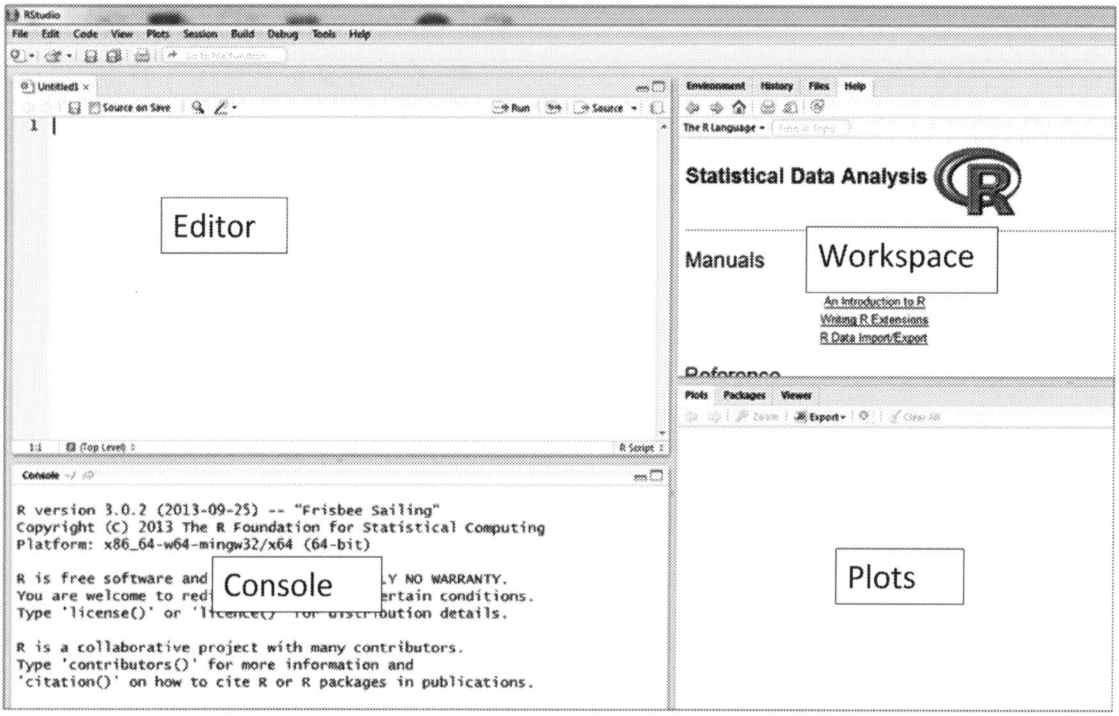

(3) Click on Tools/Global Options and select the default working directory. Alternately, you can set the path by typing ‘setwd(“path to your folder”)’ to set the working directory. Use forward slash only in the path.

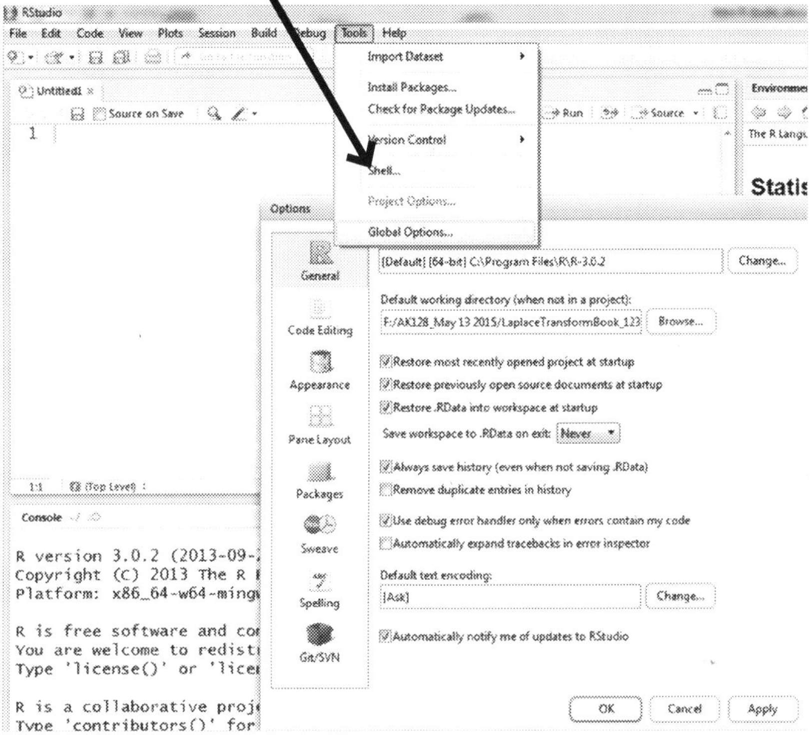

(4) If you need to install a package in RStudio that has not already been installed in R on your computer, then in the RStudio session, click on ‘Tools’ then ‘Install Packages’ and type name(s) of packages as shown below.

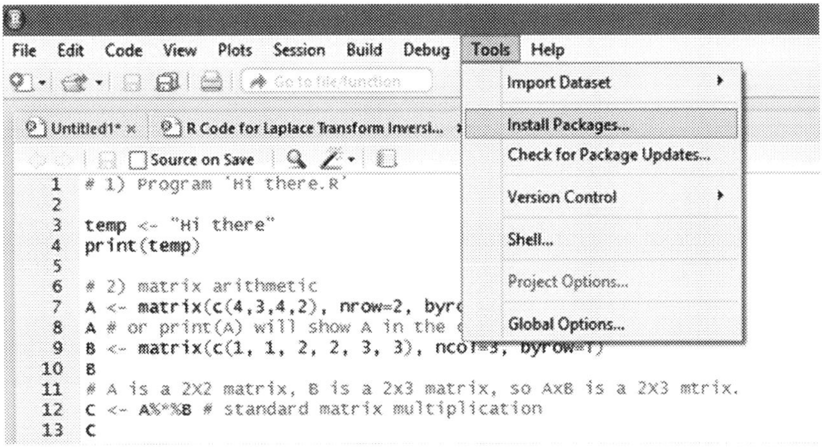

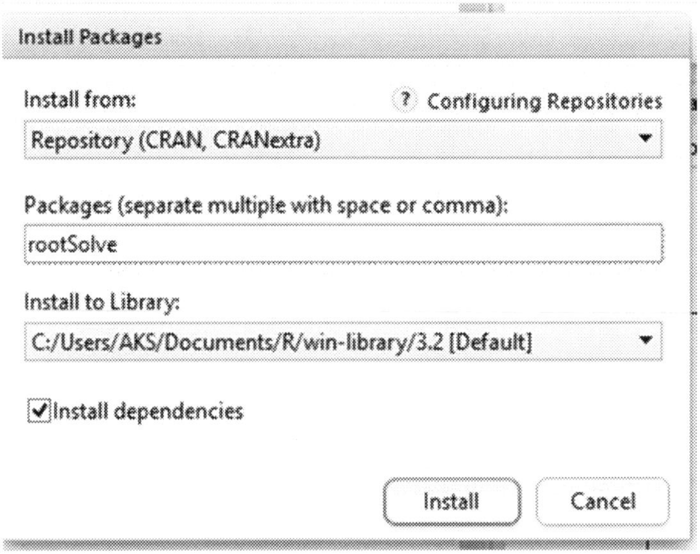

(5) 5) To run an existing R code from RStudio, click on File, then 'Open Project', and select the R-file, which will open the code in the Editor window.

Example (Solve non-linear equation(s)): See examples 3(d) and 3(e), Section 4.3

```
# R-code for solving a non-linear equation(s) in R
library(rootSolve) # the package rootSolve can be used to solve a non-linear system of equations.
# Example (a)
fun <- function (x) cos(2*x)**3
curve(fun(x),0,8)
All <- uniroot.all(fun, c(0, 8))
points(All, y = rep(0, length(All)), pch = 16, cex = 2)

# # Example (a): Solving a 3x3 non-linear system of equations

F <- function(x) {
F1 <- x[1] + x[2] + x[3]^2 -12
F2 <- x[1]^2 - x[2] + x[3] -2
F3 <- 2*x[1] - x[2]^2 + x[3] -1
result <- c(F1 = F1, F2 = F2, F3 = F3)
return(result)}
```

```
# first solution
 (ss <- multiroot(f = F, start = c(1, 1, 1)))

# second solution
 (ss <- multiroot(f = F, start = c(0, 0, 0)))
```

To run the entire code, highlight all lines in the code in the RStudio Editor window, then click 'Run' which will produce a graph in the Plot window for Example (a) and the results for Example (b) in the console.

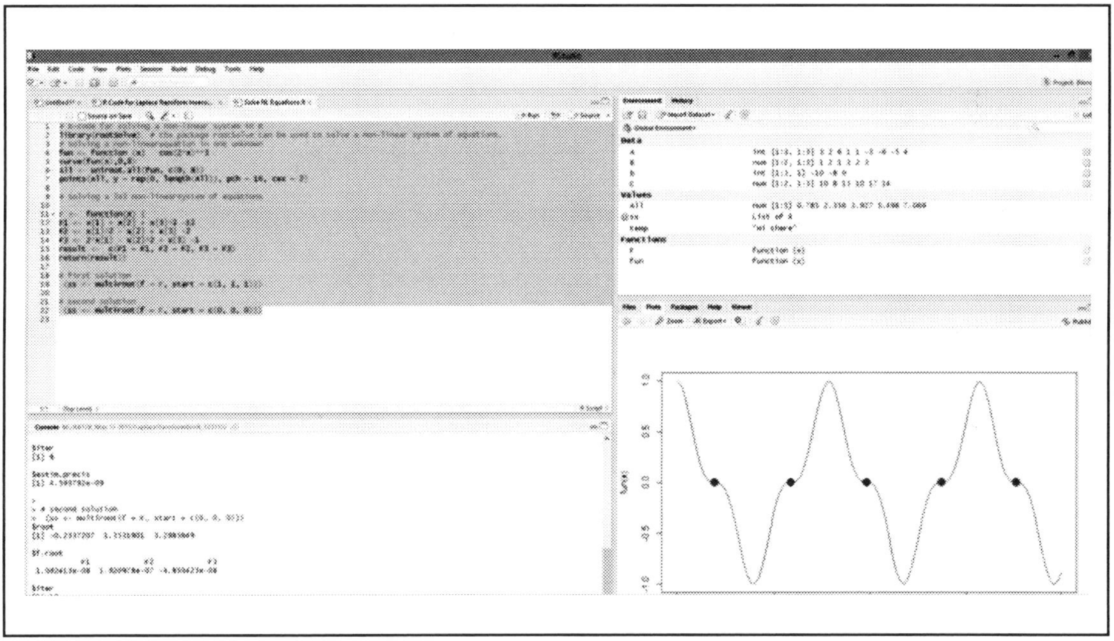

ABOUT THE AUTHORS

Dr. Rohan J. Dalpatadu is Associate Professor of Mathematics in the Department of Mathematical Sciences at University of Nevada, Las Vegas (UNLV). He earned his B.Sc. (Honors) (1974) in Mathematics from the University of Ceylon, Colombo, M.S. (1981) and Ph.D. (1986) in Mathematics from Southern Illinois University, Carbondale. He is also an Associate (1991) of the Society of Actuaries. He has taught undergraduate and graduate mathematics, statistics, and actuarial science courses. His research interests include numerical analysis, applied mathematics, applications of probability and statistics in gaming and actuarial science. His email address is dalpatad@unlv.nevada.edu.

Dr. G. S. Singh (Nov. 14, 1931 – Sep. 27, 2009) was Professor & Head of the Department of Mathematics at the DAV Postgraduate College, Lucknow, India; he retired from this position in 1989. He earned his M.Sc. and Ph.D. (1960) in Mathematics from the University of Lucknow, India. He wrote his dissertation in the field of Hydrodynamics. He has numerous publications in mathematics. He was very interested in Sanskrit, German, and French languages; he taught German and French languages at The University of Lucknow.

Dr. A. K. Singh is Professor in the Hotel Management Department at UNLV. He earned his B.Sc. (1968; Physics, Mathematics, Statistics) and M.Sc. (1970; Mathematical Statistics) from The University of Lucknow, India and Ph.D. (1977; Statistics) from Purdue University, W. Lafayette, Indiana. He has taught undergraduate and graduate statistics, mathematics and OR operations research courses in the Mathematics Department at New Mexico Tech (1978 - 1991) and the Department of Mathematical Sciences at UNLV (1991 - 2005). Since 2006, he has been teaching Mathematics of Casino Games and advanced statistics classes including Time Series Forecasting and Data Mining. His research interests include: Bayesian methods, reliability applications, business statistics, applications of statistics in hospitality, gaming and other disciplines. He has more than 90 publications in theoretical and applied statistics. He can be reached at aksingh@unlv.nevada.edu.

Printed in the United States
By Bookmasters